SECLUDED RENDEZVOUS

CAROLYN BEGG

Secluded Rendezvous

ISBN: 978-1-935125-04-4

Copyright © 2008 by Carolyn Begg

All rights reserved under International and Pan-American copyright conventions. No part of this publication may be reproduced, stored in a retrieval system or transmitted in any form or by any means, electronic, mechanical, photocopies, recording or otherwise, without the prior written consent of the author.

To order additional books, go to:
www.RP-Author.com/Begg

Printed in the United States of America

Robertson Publishing
59 N. Santa Cruz Avenue, Suite B
Los Gatos, California 95030 USA
(888) 354-5957 • www.RobertsonPublishing.com

Dedication

I would like to dedicate this book to my son Randall and grandson Josiah. You are two of the most important men in my life. You have always shown your love, hope, and trust in me. I love you both, and wish for nothing but the best for you. I am very proud of you and hope you enjoy "Secluded Rendezvous."

Forward

I pray often that I may not be disappointed in myself, this achieved, and I will not disappoint others. I have found a life here, a very spiritual one living in the mist of the beauty and peacefulness which was given to us.

The Lodge, being built among the thicket of tall trees, is only 30 feet from the shoreline. I have soft ferns, lemon grass, and moss growing all around me with many trees. The sea and other islands are my front yard; I find myself in two places the calmness, and wildness which Mother Nature provides. Some of my most pleasant hours were spent during the long wild rainstorms, where I was cozy and protected inside by the fire as thunder and lightning raced across the sky. Then I would run outside on the deck and be a part of the clouds, wind and rain. The wind swift, would howl and whip around me. The trees and sea would shift up and down, back and forth in a wild extravaganza. All of a sudden, it would be extremely silent—not a sound. The storm was over! It was as though you were completely alone in the world. Quite often I was the only one on the island, but I knew God, Jesus, and my son were always with me. I never felt alone.

Chapter 1

I am not sorry I made this journey. It is not where I want to spend my life. It is an adventure, and I feel brave and proud that I dared to undertake it. I do feel cut off from my family and friends. A part of my mind and heart are back home. The island people are nice, but it's not the same. I miss my son so...I often go out on the dock, and talk to him. I tell him I am here because it is where I have to be for now. I will return as soon as I can. I carry you Randy, in my heart. Sometimes I do not understand why God chose me to do this...or did he?

I thought this was a way to enrich our life. Build a business, keep it for five years, sell it, and return to California and live comfortably. How rewarding it would be. We never planed to be there twenty years. Where did the time go? I did return every year for three or four months each Winter. I worked and went to College and got my degree in Floral Design. I stayed with my son, or rented an apartment. However, I missed Rob. My thoughts were forever going back and forth. The drive from Canada took about sixteen hours. While I was on the island Randy and I talked every Sunday. When I was in California, Rob and I talked back and forth. When I was on the island, I became a part of the environment. The vast empty spaces, the trees, hills, wild animals, the islands, and the sea all became a part of me. I did have an intense feeling about the injustices of the world. I was not satisfied to let life be. How could I live here in this heavenly peacefulness, and my family and friends in that other world? Days marched by, hours were lost. The calendar was meaningless. Time did not matter.

Carolyn Begg

We woke with the sun, and began our day, working till dark. There was no rigid schedule with the exception of working the tides. When there was rain or snow we put on our wool sweaters and rain gear. We picked up our saws and hammers and went to work. Out there one cannot let the weather dictate your time. Living out there was an experience most people do not have in a lifetime. One cannot even begin to imagine the knowledge you accumulate on an everyday basis. You must learn to be very perceptive.

We had a 21foot aluminum correct craft. She had a 350 cubic inch Ford Mercury cruiser. We also had a 14 and a 16 Foot aluminum boat. Our main boat (and my favorite) was a 17 foot welded aluminum. It had a bar on the back for pulling logs. There was a console in the middle with a windshield and a seat for driving. It was a wonderful boat. I learned to drive them all. What fun! You have to learn how to drive in whirlpools, rapids, fog, and storms. Being 30 miles from town, you had to learn to do light maintenance on the motors, read charts, clouds, and read a compass. I loved learning all this. It was so different and new. We drew our own plans for the house. Robbie did the actual drawing. We learned how to cut trees, limb and peel them, measure and cut lumber, and of course nail and saw, and how to build. Rob knew most of this, as he was building when he met me. I had to learn how to mix concrete, pour foundations, run my chain saw, an Alaskan Mill, and a Mobile Dimension Saw. I think the most frustrating part of taking on this project was the time it took to complete the jobs. For instance, to pour the concrete, you must first make the concrete.

We bought a 1956 Chevrolet truck. It was a wonderful thing to take to the island. We also bought a Mobile Dimension Saw. The saw had a Volkswagen engine and

Secluded Rendezvous

it was portable. You set it up and brought the logs to it. We could mill 20 foot lengths with this saw. We also purchased an Alaskan Mill, which is a chainsaw with a six foot blade. One person runs the motor, and the other person is at the other end of the blade holding the guide. You learn to have a great respect for these saws. We took nails, tools, shovels, axes, rakes, smaller chainsaws, a wheelbarrow, a couple of generators, a bathtub, and anything else that would fit on the truck. We knew once we got there we had no store. We had to bring what we could. The store was a long way from home. We also brought a 16 foot trailer and a tent for company to stay in. My son, Randall, his girlfriend Kathy, and Alan, came with us to help us get started. Randy drove the truck with Alan and Kathy. Rob and I drove my car. We pulled a 21 foot Correct Craft boat with my 1970 Ford Torino. It was very hard on the car... not something I was in favor of. We arrived in Canada two days later. We took a ferry out of Horseshoe Bay in Vancouver. We had lunch on board. It was a lovely trip across the water. Two hours later, we drove off the ferry in Nanaimo. We then drove along the coast into Campbell River, it was another two hours. The drive is spectacular, the water breathtaking.

Campbell River is known for its Salmon fishing. It is an excellent location for Anglers. The fish run through Upper Discovery Passage and Seymour Narrows into Johnstown Strait. Campbell River is a small town with lots of little shops on its main street. There are a couple of small shopping centers about town. The town has a hospital, a liquor store, gift shops, grocery stores, restaurants and other shops. It's a fun little place, we did a tour before we left it. This is where we would buy our food and supplies. We drove on the ferry that was heading for Heriot Bay, on Quadra Island. The ferry is only a twelve minute ride. It stops in Quathiaski Cove. The cars drive off and go North across the other side of the island, to reach Heriot Bay. There you will find a Government Wharf and the Heriot Bay Inn. There is a small shopping mall where you can buy groceries, liquor, and other needs. Quadra Island is a wonderful island. Many people live there and commute to work in Campbell River. The homes on the island are very unique. There are many old farm houses and small farms. When we arrived at Heriot Bay, we launched our boat. Randy and I drove the car and boat trailer up the hill to a back yard. A nice couple and their family live there. They have a large yard, all lawn. The wife rents spaces for cars, trailers, and boats. It is very clean, and a safe place. We felt very relieved to find such a place, so close to our island. It takes 45 minutes by boat, to get home. Randy and I hiked back to the wharf, where the boat was loaded with gas, food, supplies, Rob, Kathy, and Pokey, our dog. We started for Rendezvous Island. Home! I had my first experience with rough water. The swells were high, and Kathy was very scared and sick. Poor thing—green was her color when we got to our island.

There are three Rendezvous Islands. The North, South, and Middle. Our ten acres were on the North. It was on the Calm Channel side. The island is 1 and 3\4 miles long, and 1\2 miles wide. The three islands are located about 130 miles northwest of Vancouver, British Columbia. North Rendezvous is the largest island of the three. The islands are situated north of Cortez Island, and south of Stuart Island. It is lying near one of the finest salmon fishing areas in the world, Campbell River, which is about 30 miles south of us. Six miles north of us is the Arron Rapids. The rapids are about one cable wide, in the narrowest part. The flood tide, from what is

Secluded Rendezvous

known locally as the Yuculta (Eucleutaw) Rapids, make it very hazardous for a vessel to pass through, except at or near slack water. In June of 1792, George Vancouver's ship was pulled through the rapids by the Indians. The Indians pulled the ships with ropes. Now a good boatman can make it, on his own. Others do not! Captain Vancouver and his men used to rendezvous around the islands. The islands were then called the Three Maria, but he changed the names to the Rendezvous.

Ten minutes from our island, heading south, is a passage called White Rock. You must line your boat up with the markers on the beach, to get the boats through. The passage is very narrow. At the end of it is a general store with liquor, books, candles, can and fresh food, bread, candy, and many other goods that people out there would need. There was also a Post Office and gas pumps. It was a Godsend for us. Two ladies were caretaking it for a man who was ill. This man passed away and the two women took it over. We became good friends with them. Wonderful people. After two years of owning it, they sold it, and moved to a better place for them to live.

On the Southeast corner of Rendezvous, there was a one-room school for many years. It was used for the settlers on our island, and local children on nearby islands. There was a large vegetable garden and a small orchard. The cherry and apple trees are still there. This homestead, on the west side of the island, is at Marks Bay. Here, in 1910, a Japanese fisherman's family built their home. It was very sunny on this side of the island. A wonderful place for a garden and fruit trees. There we discovered old rhubarb and rose bushes growing wild. The place was not called Marks Bay in the beginning, rather, Ishi Cove, after the family that lived there. A nice name, I believe.

Carolyn Begg

We pulled our boat into Marks Bay, and tied up to an old float left by the realtors. We unloaded our boat on the beach. We tucked some things behind a bush, and we all carried what we could and needed for the night. The rest we left for two days. The Marine Link would come with the truck and trailer, and the rest of our belongings. When we reached our land we walked it and decided where the trailer would go. We began clearing the area for the trailer. The land was thick with bracken. We also cleared a path from the road to the trailer. In the middle of the island was a logging road. It came up from Marks Bay (west) and ran into a road going north and south to each end of the island. Our property ran from this road east, down to the beach. We loved the ten acres we had. A small farm would be on top, and a beach on the bottom... perfect!

On the morning of our second day we all had breakfast and went walking down the road. No one spoke. It was as though we were in another world. A shivering group of trees, and further down were the sun rays leaping from behind the trees. Breathtaking! My son whispered, "Mom."

"I know son," I answered. The little birds were hopping in and out of the ferns, some eager ones flying from tree to tree. Some were chirping at us. We laughed. It was as though they were asking, are you going to live here with us? We reached the beach and the barge had not arrived. We had no way to communicate with them, so we explored the Marks Bay and Old Ishi Cove homestead. As I looked at an old shed (one room) I wondered what it was used for. We all talked about what we thought... this must have been this, and I wonder what this was. You know, I thought years from now, people will be doing this to our place. Just think of finding our old house, falling over. Oh, my! We sat

Secluded Rendezvous

on the beach and waited for the barge. The tiny sand crabs sat with us for awhile. They watched us and we watched them. Soon they were scurrying sideways to hide under the rocks. We were hysterical. We talked to the kids about our plans. My son was very excited for us and our new adventure. However, he could not hide the pain in his eyes for the separation of us-nor could I. My chest felt like it was tearing apart. I wanted to hold him and say take me back with you. I could not.

I had to do this for all of us. I wanted to be secure and have something to leave him when I passed, and not be a burden to him when I'm old, as I have seen so many doing. I thought, in five years we shall be finished with this place, sell it, and return home. We will be back with our children and they can watch us grow old. It all seemed like a fine plan at the time.

Marine Link arrived in the afternoon with the Captain, Tim. Tim owns and runs the Marine Link. He works up and down the coast. He lives in Campbell River with his wife and children. He is a very sensitive and caring man. I do not know what people out here would do without him. We hooked the trailer up to the truck, unloaded the rest of the barge, and said goodbye to Tim. We arrived at the top of the hill, parked the trailer and truck, had something to eat, and went to bed. By then it was too dark to see. We all had a busy day and were in need of sleep. The next morning we woke at sunlight and drove to our property. We put the trailer in place and began to unpack. Randy set their camp sight up on a little bluff, among the trees. It was a lovely spot. We had fun the five days the kids were with us. We worked, played, explored, and fished. We got everything set up and ready to go. At the end of five days, the kids had to return home to work. The week went so fast. What will I do without him? However, we still had time together, as Rob and I were driving them home to California.

We took our boat to Heriot Bay. Took the ferries and drove to Vancouver. We stopped at Rob's mother's and left Pokey there. We drove to California, stayed overnight, and left at five the next morning, for Canada. We drove to mom and Haakon in Vancouver. She and Haakon loved Pokey. They always had a wonderful time together. We stayed overnight and returned to Rendezvous the next day to begin our new adventure. We walked around lost that afternoon. Went for a walk to the property next door. There was an old tiny building there. We were told a girl named Judith lived there. She was with child and away for a time. She returned with a baby boy named Ian. It was so quiet, just the two of us. We got to bed early and were up at dawn, ready to get to work. We sat up our Mobile Dimension Saw at the top of our (to be driveway) next to the logging road. We knew we would be cutting close to the road and would be pulling some trees with our truck to the mill. It took most of the day to clear the area around us. We raked everything in a huge pile to be burned in the winter.

Chapter 2

The afternoon of this day, a man came walking up the road, and was very excited when he saw the sawmill. He introduced himself as Louie. He laughed and said, "I heard a couple of city folks arrived from California, and I stopped by to see if you two needed a hand." You don't *"stop by"* on the middle of an island. I knew this man would be a very nice man, and he was. "I am a tree cutter," he told us. We were happy to meet him. He had never seen a saw like this—only heard of them. He said he was born on Read Island, next to us. We all looked over the saw and Rob said we would begin to cut tomorrow.

"What are you using for wood?" Louie asked.

We looked up at the trees and replied, "We both have chainsaws, and we will cut down a tree, as we need it, and bring it to the mill".

"Oh," said Louie. "You know how to cut down a tree?"

"Well, sort of, we can do it... I think," was my answer. Louie gave us a big smile, and a few tips, small ones like—the tree can fall on you, if it has a curve in it or if the wind is blowing it your way—or, it can buck you and take your head off. Needless to say, we hired Louie to cut our first three trees. What a marvelous teacher. The guy was so impressive. He could draw a line on the ground, and fall the tree right on the line. He did this to show us how to plan where to make your cut and where to have the tree go down. This was so you didn't hit other trees with the one you were cutting. We learned a lot that day. Louie told us his dad was a tree faller, and taught him when he was six years old.

He said, "This is how I earn my living, I have been doing it all my life." I told him it was nice of him to teach us when this was his way of making a living. Of course, we hired him during the years. Sometimes a huge tree, or one that lightning hit, or one that required a cable attached to it to get it to fall right … we would hire Louie. I learned a lot from Louie, but I left the big cuttings to Rob. I cut the limbs and branches off of them after they were down. The morning after Louie left, we were up at four a.m. We could not wait to try our new saw.

We used the truck to pull the trees to the saw. This began our sawmill experience. When we made our first 2x4 we couldn't believe it! It was exactly accurate, by our measure. What a burst of pride we had, to know we cut our first board for our home. Rob said he would cut and I would carry and stack the boards. We had a nice little pile by the end of the day. We were dead tired but it was hard to stop, even for lunch, we were so excited… we continued this process for about two weeks, working from dawn to dusk. A Mobile Dimension Saw is said to be a one man saw, but it is very hard for one person to run it alone. It leaves a lot of sawdust as it cuts. We talked of using the sawdust for the outhouse and the road. So, Rob ran the saw for a few days, and I worked with the sawdust. I started pulling the bracken and stacking it on the burn pile. The bracken was taller than I was. It was kind of fun working in it. I began making rooms as I pulled. I was playing. The burn pile grew fast. The island was basically made of rock, ferns (bracken) trees, and covered with Lemon Leaf. It's used very much by florists. It is a beautiful plant. Every place you dug or raked, you would find a rock. I dug out rocks and placed them in my road to make it level. I filled the wheelbarrow with a pile of sawdust around the rocks. I believe this is how I really learned to use a shovel. Every project I worked on was a new learning experience for me. Yes, it was hard, but I loved doing it.

I was a graduate from Abraham Lincoln High School, in San Francisco. I was a fashion model for three years. I danced, one summer on television with Arthur Murray's group. I gave up my Joseph Magnin boots for my rubber gum boots. I loved my new look. No place I had ever lived, gave me the sense of achievement that this place did. Everyday you woke up to a new challenge, and you must learn, and figure out how to solve it… and do it. Rob left the saw for a few days and started chalk drawing on the ground. He dug the holes for the Redy Mix, for the foundations. We mixed the Redy Mix in a wheelbarrow and put it in the holes, along with the cut ends of our trees. The ends had been painted with preservatives. Our floors were soon to start. We returned to our milling.

Chapter 3

My brother-in-law, Edward, my niece Leslie, nephew John, niece Cindy and her husband David, are coming from California. I am so excited to see them. We took the boat to Heriot Bay to meet them all at the dock. We left early in the morning. We pulled up toward the dock and they were all waving and yelling. We were so happy to see each other. David had already parked the car next to the Torino, and they had their belongings on the dock. We loaded the people and gear and headed for home. They couldn't get over the beauty of the water and islands we passed. It was a lovely day. I was happy and thankful for the calm water. After we unloaded the boat and drove the truck up to our place, they were anxious to see the land, and we were anxious to tell them of our plans. They couldn't believe the pile of wood and clearing we had done. David is a construction man and will be helping us with the house. What a blessing! We all took a tour of the land. They loved the quiet and beauty of it all. We had a nice evening drinking wine with a nice dinner. We were in bed early.

Everyone was up early and ready for work. The Mobile Dimension Saw was down the trail in a group of trees, we had some lumber cut ready to go. The trail was steep. This made it hard to carry the lumber up. We carried one at a time up and down the trail. Around noon, I was carrying a board up, and Edward was passing me, as we passed I thought, he doesn't look good. On the next trip I stopped and asked him if he felt all right, he said, "I'm feeling fine, Cindy just asked me the same thing what's wrong with you girls anyway?"

Carolyn Begg

Edward was really sweating, and his breathing was short and very fast, he was white. I didn't want to upset him, so on the next trip I said, "Let's all break for lunch." I talked to Ed. He said he was having some gas pains in his chest.

Rob said, "We have enough wood here to start, so David and I will begin the walls." We took a rest, while they worked on the kitchen and pantry. The kitchen was 16x20. The pantry was to be 6x12. They were there two nights. In the morning they said they were leaving. We were all very worried about Edward. He would not admit to not feeling well, but we all knew he was very ill. We got in the boat and took them to Hariot Bay to get their car, and they left. I knew he needed a doctor. I was happy when the car left. I knew I would miss them, but the island was not the place to be if you are seriously ill. Rob and I returned to the island.

Pokey spent the time looking for everyone. He got a lot of attention from them. He's such a good dog. We had lunch, filled our tea cups, and talked about the wonderful time we had with the family and how wonderful it was to have all that loving help. We worried about Edward. We had no phone. We worked on the house for a few hours. We drank more tea and took it to the trailer, and talked about our working plans. There was so much to do. It was hard because everything came first in our heads. Everything was left till morning. After breakfast, Rob said, "I'm building a shower and an outside toilet." I shouted "Yea!"

Rob went about 20 feet, in a small group of Hemlocks, about 12 feet high. He marked the spot where the shower would go. He took the loppers, shovel and started a path from the edge of my road, across from the trailer. I got the rake and pulled bracken from the trailer, about six feet out from the door and the length

Secluded Rendezvous

of the trailer. I then walked up to the rocks where the moss grew. It was about three inches thick. If you carefully slid your hand under it, working the roots with your fingers, you could lift up long sheets of it. I carried these sheets to the trailer and laid them on the ground. We had a moss lawn. It was nice and worked well. Rob cut out small trees the size of a pallet and laid the pallet down. He placed a Hemlock tree limb at the head of the pallet, and put a couple of hooks on it for the shower bag to hang on. Then he made a bench to sit on. The shower was a plastic bag with a hose attached and a shower head. It was the kind that is used on sail boats. It hangs in the sun and gets warm. What a delight taking a shower in the mist of the Hemlock trees. It was even delightful in the rain and snow. The shower was cut shorter in the snow. Away from the shower was the toilet. This was a deep square hole with a large box over it with a hole cut in. The box was the height of a regular toilet. The cut hole had a regular toilet seat on it. The lid could be put up or down. There was a small basket for rolls of paper. One day as I was sitting there, a deer came out of the trees about three feet from me. I sat very still. She tried to outstare me. She wasn't going to leave me till she found out what I was doing, and I wasn't going to tell her! We were there a long time. I lifted my arm and she ran away.

The weeks rolled by, and we continued to work on the house. One morning we heard a hello from the road, and two men came in from Read Island. They said they had one room schoolhouse. Fourteen children, grade one to seven, and no teacher. Rob got very excited and I was torn. The children needed a teacher and we needed to work on our life getting our house together. Couldn't they find another teacher? I told myself I was being selfish. Rob went to work for the school. I stayed in the trailer. At that time I was the only one on the island. I stayed inside all day, painting with my oils I wrote letters, I read. We were told there were bears and cougars. I had no phone. There were no boats at the dock. The police knew of a group of people one had to be careful of. I was here alone… for how long, I asked Rob.

He said he didn't know, however, they needed him. One day I asked Jesus for help. I took a long walk. At first I waited for a cougar to leap out at me, or a bear to see how fast I could run. I prayed all along the way. Then I said, "Jesus, you didn't bring me all this way, away from my family, to be killed by a man, or some animal. I'm sorry I've been so selfish, you and God have been with me all my life and I'm okay now."

Robbie went out of the trailer the next morning. He had his gun boots and rain gear, lunch and thermos in hand and headed down the road to the boat. It was very dark out. I sat in the trailer drinking my tea with Pokey. How long do I do this? My answer came. You can sit here all day or get to work? I got to work! Pokey watched while I raked the whole yard. I pulled out piles of bracken. Pokey and I walked to the island; logging road (where our property began) and I began to work on the road again. I didn't leave a weed in the ground. I raked and stacked the bracken and weeds. I knew I would be doing this till the road was finished. I was cleaning a 14 foot wide by 200 feet long area. The wheelbarrow, rake, and I became fast friends. Pokey was afraid of the wheelbarrow at first, later he would jump in and wait for me as if to say, come on let's get going. I worked on the road for a couple of weeks. Pokey and I finished it. I had a Hasavarna chain saw and cut up long pieces of small trees for firewood. I also took the wheelbarrow up the road and picked up medium size rocks and lined the sides of my road to the house. There was to be a 20x20 foot space for the front yard. I made a path to where the front stairs and deck would be. The path was lined with rocks also. I felt very fortunate to have all the rocks close to work with.

Chapter 4

It was 11 o'clock at night, a very cold and dark night. We were sitting in our truck on the beach. We had been waiting since nine, when the barge was to arrive. The Southeast winds were blowing, and reaching ninety miles an hour. We were worried about Tim. Tim was the captain and owner of the Marine Link barge. We were warm with our dog, Pokey, in our 1956 Chevy, as we had to keep it running because of the low battery. We did not hear the hum of the barge as it rounded the end of the island. All of a sudden the light was blinding on the water. Like a huge, dark, monster, it seem to rise out of the water. The green lights like eyes blinking, as it grew closer, larger. There was a blast of the horn. A very large spot light danced across the beach and on to the truck. Slower and slower it came while it tried to stop. It's tremendous weight was pushing forward building waves on to the beach.

We ran out toward the barge and the Captain yelled, "I can't come in close, the wind will blow me on the rocks. You'll have to wade out to me." We were in the ice water up to our knees. Tim dropped anchor and began handing us the supplies. The anchor brought the barge a little closer. The first things off were sheets of foam. We were running from the ice water to the beach, when the pile we had on the shore began to lift and blow across the sand. I ran after them, as they can break into pieces like glass. I would make a small pile and put a large rock on top. Some broke in half. I stayed on the beach and started putting things on the truck.

Rob yelled. "Just put them on, I'm going to tie everything down." I was needed in the water. We had to work fast. The wind was picking up, turning the barge sideways. Tim was handing to Rob, Rob was handing to me. We had a full load and Tim was pulling up the anchor and heading out. He left the spotlight on us, as Rob was tying the load down. We had cans of paint, boxes of nails, roofing, rolls of plastic, insulation, tarps, tools, 4x8 sheets of pressed board and much more. The tying of ropes was just starting when Tim made his turn and the light was off. It was pitch black out. Rob had to feel the ropes in the dark. We got in the truck and it had died and would not start.

Rob said, "You and Pokey stay here and I'll walk home and bring back a can of gas and a battery."

Pokey and I sat in the truck. It was very cold. We waited for a very long time, three hours, or longer. I thought to go look for him—maybe he fell and was laying somewhere hurt. My common sense told me to stay in the truck. I could not see my hand in front of me. About an hour later I heard water lapping under the truck. To my horror, I knew the tide was coming in. I got out and the water was over my feet. I got back in the truck and was making a plan to leave, or how much longer to wait. I heard Rob calling me. He found the truck and used a few choice words at the sea. He replaced the battery, put the gas in, and I drove slowly out with Rob pushing from behind. Ten minutes longer and we would not have been able to get out. Rob drove up the hill, swerved around a fallen tree, and we landed in a ditch. Rob was burying the wheels in the dirt trying to get out. I told him let's leave it till morning. We can see then and we'll bring a shovel and a come-a-long.

Rob said, "I will try one more time."

I said, "God. Why aren't you helping us? We have been through enough today."

Rob opened my door and said, "We are going back to California this is all to hard, the truck is empty."

Secluded Rendezvous

I said, "We didn't come all this way to turn around and go back to California. How sure are you in this dark—how can it be empty we just loaded it."

He responded with, "When we landed in the ditch, the ropes came undone, and the truck is empty."

I said, "Try to start it once again—go really slow." The truck drove out! I said, "Thank you God. Why did I wait so long to ask for your help?"

The next morning, Judith, Rob, and I came back to reload the truck. We drove it home and unloaded it. We were happy! A few days later everything was in place and I was caught up with my bracken. I decided to play carpenter during the day while Rob was at school. I had asked him to show me where the bedroom wall was to be. We had walked it out, and I had placed sticks on the ground for markers. I didn't tell him I was going to do this. As soon as he left, I got my hammer out and went to work. I placed some 2x4s on the ground, 16 in. on center, and nailed them to a 20 foot 2x4. I was building our bedroom wall. We had some books to help me. When he came home we put it up and it fit. Everyday after school, we worked on the other walls and the framing was done on the bedroom. I asked him about the sheetrock that was stacked in the kitchen and he told me how to put it up. The next morning I stood in front of the stack, slid a sheet off, and wondered how to lift it and carry it. I set it on the foot of my boot and leaned back and carried it across the room. I only did this twice.

The next day we went to Quadra Island to pick up Bobbe and Joe. Bobbe worked for me at Skyline College while she was going to school. She was seventeen years old. We have been dear friends since. I love Bobbe. We were so happy to see each other, it had been a long time. The four of us toured around Quadra's shops and had lunch. The trip home was lovely. Bobbe and Joe both enjoy the outdoors. They didn't want to sleep in the trailer. They wanted to sleep in our new house. Finished or not. We made a bed for them on a stack of sheetrock. And they were happy! They wanted to help us. We wanted to take them fishing but they were both working at home, and only could take off a long weekend. Two days later we took them back to Quadra, but not before Joe and Rob hung the rest of the sheetrock. They hung it, and Bobbe and I gave the orders. I sure wish they could have stayed longer. Forever would have been nice. I missed her so...

Chapter 5

I wanted to work outside for a while. I told Rob I wanted to build a fishpond with a bench on one side. He laughed and said, "Carolyn, you can't dig a hole and fill it with cement." He went to school, and I built a pond. I dug a circle four feet round and two feet deep in the middle. I shoveled a couple of inches of sand in it and up the sides. I packed it tight. The next day I mixed the cement and using a trowel, smoothed it in. I let it set for a week and painted it a light blue. I filled it with water and it worked! A few days later I heard the sound of a frog. I ran outside and there was a frog in my pond. I could not believe it. I put a rock in the middle, for him to climb on, and named him Freddie. I had no way of knowing if he was a Fred. However, a week later there was a Freda. I may have had their names on the wrong frog, but they didn't seem to mind. They were darling. Pokey thought they were great. He sat for hours watching them. Rob built a bench for the pond and I was happy.

Rob was doing okay at school; but it was not as fun as he thought it would be. Some of his students lived on the island of Read, some of them came by boat. One 16 foot skiff was driven by a 10 year old. She was responsible for herself and three other children. She came across Hoskyn channel in the wind, rain, sleet and snow. The channel itself can be wild, and in the storms it's dangerous. Rob would go in one hour early so he would have the wood stove heating the room for the children when they arrived. He would have them remove their coats and shoes to dry out. Poor things were frightened. They would be talking all at once about the waves hitting them and the water in the boat. The living conditions for some of the children wasn't the best. A friend of the parents moved out to their area. They wanted him to teach. We were so happy to get back to building our home. I had built a entrance room, with a laundry space on one end. I framed an entry door (off the front deck) and there was a window off the side of the door—so high you could not see out. Remember you build on the ground then put it up. Does that help for the error? I think not. Well, now Rob has time to fix it. I am happy.

Chapter 6

We were going to put in a garden. There were lots of deer on the island, very friendly deer. Judith told us you need a fence around your garden ten feet high. This was off by a few feet—we later learned. We heard of a net loft on Quadra Island where fishing boats bring their old nets. Most of them are very heavy and large. They gave them to the Island people for fences. We went there and brought these nets home. We put them all around the garden. Robbie sunk posts in the ground and made a gate. It was wonderful. We were working on the house and garden. Both were in need.

One morning, while sitting outside in my rocking chair waiting for the glorious sun to show its face, I thought, we need to go to Campbell River and get a phone. When Robbie got up I said, "Let's go in and get a phone." We made the trip, bought a phone, had lunch, and returned home. Rob set it up and after dinner, we called Randy. It was so nice to hear his voice. I was sitting in my chair under the trees, in the middle of an island, listening to the voice of my son. What an absolute joy! He was so happy to have a number where we could be reached. It was a comfort for all of us. We called my sisters Arlene and Barbara. We called dad and Dee, mom and Haakon, Bobbe and Joe, and Rob's girls. We were on the phone all afternoon. What fun. Mom and Haakon wanted to come for a visit. We set a date to meet them at Campbell River.

The weather was very calm the day we set out in our boat to meet them. You can go around Quadra Island to Campbell River, but only in the calm water.

Carolyn Begg

We usually take the ferry over to Campbell River. They were waiting for us and it was nice to see each other. All were happy and filled with love. Pokey was delighted to see Haakon. They enjoyed the trip home. Haakon and mom's family are from Norway. My mother-in-law is very dear to me. She's such a lady and I love her very much. Haakon is her second husband and they are very fun and happy, we enjoy them. The water was beautiful calm and quiet. It was a nice trip. We docked at Marks Bay then drove the truck up the hill. Mom, Rob, and Pokey were in the front. Haakon and I sat in the back with the food and luggage. We laughed all the way. We were hitting all the bumps in the road. We were grabbing food and boxes as we rounded the corners. I was laughing at the look of surprise on Haakon's face, and he was laughing out of pure fun. The truck drives up the hill to the middle of the Island. The road is rough and steep. At one point you must floor it and slide up around the turn or you will not make it. Haakon held on and just smiled at this one. No laughing. We reached our driveway and there was a trailer, tent, and a house well on it's way to being built.

Mom said, "lovely!"

Haakon said, "All the way here I thought, what is it going to be like, it's wonderful a home in the middle of the island. Someone will say, how did it get here?" It was a long day for them. We got them settled. Haakon and Rob had a drink or two while we all took a walk down the road. We had an early dinner and were in bed early. They loved it all, and couldn't believe the work we had done.

We were all up early. We had a good breakfast and went outside. We sat on the porch and talked for a couple of hours. It was nice. Mom brought all sorts of food. Haakon brought all kinds of liquor. Mom and I put everything in the pantry and went back outside. I told them I was going to work on the wood pile. Rob wanted to start a shed for it.

Mom and Haakon got up and asked, "What can we do?" I got my chainsaw going and started cutting the small branches. Mom and Haakon raked the yard and picked up small pieces of wood to start the stove with. Rob got four posts in the ground and then put a roof on top of his cross pieces and we had a nice wood shed. Mom would disappear from time to time and return on the front porch, and call out "tea is ready." There were also Norwegian cookies, and so good! She brought dozens, so she could leave them when they went home. What a treat. We had put up our tent before they came, for Rob and I to sleep in, next to the trailer. Of course, they wanted to "camp out" and wanted us to have the trailer. The weather was so nice. It was fun for them. We had lots of padding under the bed.

It was a golden morning. We all piled in the truck and went for a tour of the island. We had a fun time. We drove to every place we could get to. They had an adventure and loved it. We returned home, had an early dinner and enjoyed hearing Haakon's tales of fishing in the Bering Sea. We had a few drinks, so we were in the laughing mood, and we did laugh half the night. We did this almost every night. We loved being together. They were with us for ten days and it went fast.

One morning we packed them up and took them back to Campbell River. On the ferry over we all had tears. We watched them drive out of Campbell River and knew they would be missed. We had lots of shopping to do. We bought everything we needed, and had lunch in our favorite restaurant. We got the ferry, drove to Heriot Bay, to the dock, and loaded the boat.

Chapter 7

I took the car up to the hill to park it, Rob filled the gas tank, and we started for home. As always, we loved the ride home. The realtor had built sort of a dock, and when we got home it was under water. There was a dock on the North side of the island that belonged to Harold. Harold owned a log cabin and lived alone. He had told us when we first bought our property to use his dock. We sure accepted his offer this day, and were grateful. We had a pack sack full. We had a battery, full gas cans, food, tomato plants, and squash. Judith came down with her red Volkswagen, she had heard our boat on the North side and knew the dock was under and we would need help. The truck couldn't go down to Harold's dock. Ian sat on Judith's lap and I sat on Rob. We drove to our place. I stayed with Ian, and Rob and Judith drove back and forth—six trips! Judith asked us to come home with them for dinner. They are building a house in the middle of a big yard. There was a drive through gate and road going down the side to the house. Their plan is to have a huge garden and fruit trees and a fenced in pen somewhere in the yard for goats and chickens. They plan to "live off the land." We had a nice dinner from her wood stove. We did her dishes using her hand water pump, built over her sink. Fun house. Candles, of course. They will be using a propane light on the ceiling later. We left just after coffee and apple pie. Judith is a wonderful baker. After walking home with a flashlight in hand, we put all our things away and got in bed at 11:30. A long day!

The next morning Rob said, "Let's go down to Marks Bay and build a buoy." So we made a 2x4 box, nailed it to the top of two driftwood logs, placed an old heavy anchor in the middle with steel rebar, and rocks around. We then mixed cement in the wheelbarrow with rainwater and poured it in the box. It looked wonderful. The next day we went down to Mark's Bay early to see it, and it was sitting under the water. The tide was now high. Rob dove under the freezing water with a longer rope and the buoy came up—we were happy!

The kids were coming the next day; Randy, Kathy, and Kathy's children. I was so excited I could not sleep. Neither could Rob so we talked most of the night. The truck was sitting at the dock. It wouldn't start last night so we left it there. We walked down in the morning to the dock. I let the truck know I was unhappy with it, and climbed in the boat. The person driving gets in first and starts the motor. The other person unties the rope and hops in. We were on our way to get Randy, Kathy, and the kids. Her kids were beautiful children, an eight year old boy, and a five year old girl. They were to meet us at Quadra. It was a fourty-five minute drive there, and I never passed another boat. It was early, and so beautiful driving in between the Islands. Some of the rocks are sheer cliffs down into the water, some very clean and grey, and others with moss growing deep into the cracks. The water was a deep purple-blue, and the wake from the Motor, silver and white. When it's quiet and still, I can feel God's presence. It's the most powerful feeling a person can have when driving the boat. You stop in the middle of the motor noise, and shut off the engine and look around you and listen to silence… and feel God around you. I pulled up to the dock. They were all there. I cried. It was so marvelous to see them standing on the dock all in a row.

One of the kids yelled, "Carolyn's driving the boat!" We cried, hugged, kissed, and laughed all at the same

Secluded Rendezvous

time. We packed their things in the boat and walked up the ramp to the Inn to have something to eat. Randy had taken their car up earlier, and parked it next to the Torino. We had something to eat. They told us about their trip to Vancouver, and really loved the ferry ride. They couldn't wait to get in our boat and see the island. We finished eating and walked down the ramp and into the boat.

Kathy said, "I can't believe you can leave all your things in the boat and no one touches them." I answered it took me a while to get used to too. We had a beautiful trip home. You sit back and look to both sides of the islands. These islands are filled with Salal, Bracken, Elderberries, Oregon grape, Huckleberry, and many creeping plants. There are Cedar, Hemlock, Maple, and Alder trees. Other evergreen baby trees are pushing their way up to meet the sun. Alana, Kathy's daughter, came to my side and I sat her next to me.

She said, "It is so pretty Carolyn." We talked about the growth on the islands. How the wind and birds carry the seeds to spread new life on the islands. She said, "I want to live here."

Her brother yelled above the motor, "Not me it is too far and there are no stores or houses out here." We laughed.

I said to Alana, "While you're here visiting, fill all the beautiful things you see inside you, in your heart, and in your head, and you can carry them home with you. Whenever you want or need them, they will be there for you." We arrived home, tied the boat to the dock, and walked up the hill. Everyone was carrying food, clothes, Randy's tent—the kids were great with this. It was a good look at real life on an island. We arrived at our place and put everything away. Randy set up their tent in the same place as before.

Carolyn Begg

The next morning, after breakfast, Rob left for Campbell River. He was going to meet Tim. He was coming back with him on the barge. Before he left he said, "Carolyn you and Randy bring up water from the well below."

I said, "What if he can't get the truck started?"

"He will," said Rob. Randy worked on the truck. He got it running and drove us down the hill almost to Marks Bay. There along the side of the road, was a well. Some properties had a well on them. Ours did not at the time. We had a 50 gallon drum on the back of the truck. One end of a hose went into the drum, and the other end went into the well. The children looked in wonderment as Randy siphoned the hose. The drum was filling and there was a shout from all of us. It's a funny feeling to be out in the middle of the island and be told, "there is no water." We had to find some. Everyday, I learned more. There is nothing you cannot do. If all else fails, you pray for help and it will always come. It doesn't always happen your way—but it will happen. Randy drove us, and the water home. The children and Randy helped me put some vegetables in the garden. We carefully watered them with a can. The children wondered why we didn't get water from the ocean. We talked about the salt in the water. This place was a wonderful place for children to learn. One thing being, make sure you see a well, before you buy property. We were told, "Oh, yes there is water." We took a ride to the south beach, where Rob would come in on the barge. On our way down the road, we found it to be washed out in one place. We drove back home, got a pick and shovel, put them in the truck.

Randy said, "Mom before we fix the road, I want to bleed the brakes on this truck. Get your rain gear on, we're going under the truck. Before I leave here, I

Secluded Rendezvous

want to teach you how to change the spark plugs, oil, bleed the brakes, and a few other things. I want to know you are as safe as you can be out here, when I am back home."

We bled the breaks, and were on our way to work on the road. We cleared out a big hole, then filled it with three logs Randy had chain-sawed. Then we filled in around the logs with rocks and dirt. We drove over it and down to the beach. The sea was very rough and Rob and Tim were nowhere to be seen. We finally reached the barge, by phone.

Tim answered, "Rob is here, and the barge will be there about 11:00 P.M., it's very rough and we could be later." We drove home, had dinner and returned at ten. We stayed there till 1:00 a.m. We went back home thinking they had tied up somewhere for the night. Rob woke us up at 3:00 a.m. Randy and Rob drove back to the beach. There were two men working with Tim, and they helped unload the barge. Rob and Randy drove a load up with the truck. I fixed breakfast while they unloaded. We ate and had our coffee. We three went back to the beach, and loaded the truck again. We made two trips, and the road was great.

Kathy was never with us on these trips. She had a very serious accident and almost lost her leg. She was a trouper to even make the trip to the island. She and the kids stayed at the camp. We did all go out for a ride in the truck, but not far. It was hard for her, as the logging road was very bumpy. She was afraid of the boat so we didn't go on the water. She never complained, but it couldn't have been fun for her. Their week was over. We took them back to Heriot bay. Randy and I walked up the hill to get his car, we said goodbye there. We got back to the dock loaded the car, and they drove off. I watched his car disappear from the dock, as I sat on the covered engine in our boat.

The four arms waving out the windows, the children yelling, "Thank you, love you." I started to cry and thought—*When will I see him again? What if they get hurt on the way home? What if I die here, and we never see each other again. What if, what if. Then I asked myself, where is your faith?*

I realized Rob had been yelling, "Carolyn stop your crying." I told him I was sorry, but why didn't he hold me instead of yelling. I remembered Kathy. When she got out of the boat, she said, "I'll never come back to this place again." And she never did.

Chapter 8

We got back home, to the trailer. We set up the generator, took showers, and plugged in the hair dryer. We had a good laugh, using a hair dryer in the middle of the woods. We went for a walk around the island and talked about what we were going to do in the morning. I wrote letters to home. We had lunch and worked in the yard for a while. We were tired and I missed the kids. We took a bottle of wine, Pokey, and went on the bluff. You can see the water from there, and watch the boats go by. The bluff is covered with moss. It's beautiful there. No one knows it's there and it is just behind the house, high up. I thought everyone on this island must have their own "Special Place."

The next morning we were up early as Harold was taking us fishing. He was showing us some hot spots. We went to Frances Bay. It was across the water from us. A lovely place, with a sandy beach. We had to be careful of the bears, as they liked the beach, too. We never took food there. I caught two salmon. We were all excited. I told Harold when he wants a fish, I would be happy to get one for him. We had heard Harold was a very good fisherman—just unlucky (or being nice to me) that day. We brought the fish home and Harold stayed for dinner. He was such a nice man.

The next afternoon we drove the boat to the post office on Read Island. There was a letter from mom. They wanted us to call, as they wanted to come for a visit. I was so happy, but worried. Could she handle the island living? She was such a city person, and such a lady. We phoned them and I told her about the shower, toilet, etc, She said, "Oh I know I'll love it."

Carolyn Begg

So we made arrangements to pick them up in Campbell River at the airport. The date was set, and we picked them up. They loved the trip back to the island. The water was so smooth. We drove very slow and enjoyed the scenery.

Mom turned to me and said, "I've been so worried about you honey, I'm so glad we came to see all this. I couldn't imagine the stories of the boat, the ferries, the car up the hill parked, the truck up a logging road. I can't tell you how lovely it all is. The water, all the little islands, this road you built, my, all the things you kids have done it is unbelievable." None of us had ever seen her so excited.

Later Mitch said, "I thought your mother would hate it here, and be afraid, she loves it!" So began their vacation. We showed them our trailer, house, garden, shower, and toilet. We all had a good laugh at the last two, but not mom. She thought they were great. We took them fishing. Mom caught her first fish, and let out a yell for joy! I was so happy for her. That night, we ate her fish, and played cards. My mother was the best poker player around. She used to go to the poker room in Reno, and play cards with the men. They told me once, your mother is a joy to have at our table—and she is a darn good player too.

One morning, on their second day, she stepped down on the side of a step and fell. She hit her eye on a rock. I ran to the trailer for ice, Rob ran to the phone, we all piled in the truck, and drove down to Harold's dock. The plane was there when we arrived. It takes twelve minutes for the plane to get there. The pilot called for a cab to be there when we landed. We took her to the hospital. The doctor said she had to go home to her eye doctor, her eye looked bad. We drove to the airport. She was so sad she had to leave, but could not cry because of her eye. She told me later how hard that was. She had drops put in her eye, and a patch on it. Poor thing. Mitch called after she had seen the doctor. She was to go to him everyday for a week. One side of her face was black and blue and twice its normal size. Her eye would be okay, but it would take some time. We called her almost every day her first week. She said she was glad it happened, because she got to know how fast I could get to the hospital if something should happen to us. What a sweetheart..

One morning Rob got up and said, "Let's go to Vancouver for a couple of days, and look for a stove for the house." We went to Heriot Bay, left the boat at the dock, got in the car, took the ferry to Campbell River, drove to Nanaimo, got on the ferry to Horseshoe Bay in Vancouver, and drove to Mom and Haakons. She and Haakon were as happy to see us as we were to see them. We didn't call because we wanted to surprise them... and we did! We had a lovely dinner and sat on the patio. We listened to Haakon tell his fishing stories. It is always fun and interesting. He goes to the Bering Sea for months at a time, stays home until the next trip. He loves it. He never tells us about the dangerous trips, but we hear them from other people. When they had a big run on the fish they had to keep fishing, some days at a time. It was very hard work. We were up early, the next morning. Haakon had breakfast on the table. Rob and I left after the wonderful breakfast and went looking for a wood stove. We had heard of a store that sold just what we wanted. This stove had pipes built in to give hot water to a water tank. It also had an oven below. I fell in love with this stove. It was chrome, blue, and black. We bought it. We had it shipped to Tim's warehouse in Campbell River. We returned to mom and Haakons. We all went out to the "Son's of Norway" dinner and as always, had a fabulous time. Of course they wanted to hear about our adventure. Some thought it was great! Others thought what are you doing! "Carolyn with a chainsaw—no." One man said we couldn't do it.

We stayed with mom overnight and left in the morning. We were anxious to get home, but first must stop in Campbell River, to see Tim and buy cement and tiles for the kitchen. These were going under the stove. We found what we were looking for and headed for home. Our home was all framed in now, and sheetrocked. There was a front porch with a door that led to an entry room. There was a coat rack that Robbie built. There was a space for a wringer washer and two sinks. From this entry room you walked into a 20x16 foot kitchen. To the left of where the stove would be, was a door leading to a bathroom. The bathroom would have a hot water tank. When you entered the kitchen and turned right, you went into a 20x20 bedroom. It had two closets across one wall. They were very long and worked well. We had a sunken Library. A door went out to a small deck with a path that led to the outhouse. We also had a pantry in the kitchen. All of this was not finished yet, but we were working on it. We had put our food and supplies away and had a cup of tea and some fruit on our front deck. We bought two rocking chairs, we sat them outside, and we used these often. It was a nice place to look at the trees and all around us. We were looking down the road to our trailer—our first home. We could see our garden from there. It was peaceful, a place to plan. Pokey loved to be beside us.

In the morning Rob said, "I'll start the tile. In two weeks the stove should be here. I want to have the tile and rock in when it arrives." We had collected rock from a small rock pile. The realtor had dynamited the logging road from one side. It was like finding a pot of gold to us. We just picked out the rocks we wanted and put them on the back of the truck.

The next day, Robbie began to work on the tile. This was a job for one person and he was best at it. I went out to the garden, with the plants we got from a nursery along the highway on our way back from Vancouver. They have winter Pansies here that are used in the snow. You plant them in the begging of winter. They can be covered with snow. When you brush them off to uncover them, their little faces are looking at you. It was a warm day. The garden was my comfort place. I love to put my hands in the soil, and feel the warmth or cold of the earth. It connects me to nature, and brings my son deep in my soul. A garden is a place I can go, pull weeds, dig the soil, love, and make our earth beautiful like God planned for us. I believe God wants us to live our own lives this way. To take care of each other like our gardens, and we will burst with love and bloom and grow. I planted the plants, and watered and went to the house to fix lunch. We had canned soup, crackers, and tea. Coming from town we had meat and would have meatloaf with potatoes and carrots for dinner. This would last us for three nights. Rob had the tile on the floor. It looked wonderful. After we stood and admired it we went outside to get the rocks. The wall behind the stove would be solid rock. One side would be in the kitchen, and the other side in the bathroom. When the stove was lit, it would heat the rocks that would heat the bathroom. Two days later the tiles were dry and Rob built the wall. The whole thing was beautiful.

A few hours later two men came up to the house to meet us. They were brothers. Their names were Peter and Deiter, very nice men, from the south end of the island. They said they just bought twenty acres. One brother, Peter and his wife Edde, were from

Carolyn Begg

Germany. Dieter and his family lived in a small town near Vancouver called Ladner. They told us the property belonged to Peter, Edde, and another partner also from Germany. These two families would be building and coming every Summer. Dieter and Brigita and family would be there off and on. It would be nice to have families here in the Summer. They asked what we were doing for water. We told them about our run with the truck down the hill.

The youngest brother said, "I can teach you how to witch for water." We had never seen this done and he said, "I will come tomorrow and teach you."

I could not wait for the sun to come out so we could see how to do this. Just after breakfast, he arrived. He was carrying two coat hangers. He said, "I'm ready to teach you to witch. Come, we will take a walk in the ferns." He unwound the hangers and made a handle on the two ends so it looked like a tricycle's handlebars. He held the two ends in each hand and slowly walked around. All of a sudden his hands began to shake and the stem pointing downward toward the ground was really shaking.

We took Dieter in for coffee and I asked him if I could do it. He asked which of us had the most electricity in their body.

Rob said, "Carolyn has lots of electricity in her body." We talked about where we would look for water in the morning. Dieter went back home. He and Peter had a lot of work to do. Rob and I took the rest of the day off. Dieter told us it was best to look for water early in the morning. Rob and I sat on the deck and laughed about all of this. We saw Dieter do it, but wondered if we believed it really worked.

We were up early all excited to see if I could find water. We took the hangers and walked down our

Secluded Rendezvous

property. We were in disbelief, so I told Rob to put my scarf around my eyes and tie it in the back and lead me around. We walked a while and my hands began to shake. I took off the scarf and sure enough, we were walking over the same spot where Dieter walked.

Rob said, "This seems easy for you, so let's try on the side of our road. If you hit water there, it will be easy for a backhoe to come in and dig the well." We went up the road and I found water. We used the scarf to cover my eyes. I wanted to be sure. We would go to town to arrange for a backhoe and meet with Tim to bring the driver out.

The next day we went to Campbell River. We met Tim and told him our plan. He said he would make all the arrangements and let us know. That was easy. We had some things to do in town anyway so it worked out well. We went to lunch and talked about Tim and what a nice man he was. He is always ready to be helpful as to what is best for you. It isn't easy to drive a huge barge out here. Sometimes we see him out there fighting a storm when the weather is bad and he should have stayed in the harbor. While we were in Campbell River we went to order a Mobile Radio phone. It would take a week and the cost was $900.00 dollars. We had heard it was $600.00, from the local people. They laughed and said we were the new Americans. Who knows? We needed it so we ordered it. It would go in the boat. We drove to Courtney and ordered some Styrofoam to put under docks and a chain for making a buoy to tie boats to. We stayed overnight, had a nice dinner, got our everyday supplies, did the laundry, and headed home in the morning. Every time we went to town I would say lets get our haircut, and shop for a few things I need, like jars for canning, garden tools, there was never enough time. Thus began my cutting our hair. Robs was easy. Mine was a trick, but I mastered it.

Carolyn Begg

When we pulled into Marks Bay, there were people there camping, a man and his wife, and two boys. They told us they were from Vancouver. We introduced ourselves. Their names were John and Carol and their boys Wesley and Jordan. They had bought 10 acres right in Marks Bay. We were all happy to meet. They were nice and friendly. They asked about living on the island year round. They bought their place for summer vacations. They were working and younger than us. We went home and read our books. We are both readers and don't take time to do it too often. It is nice when we do. We had rain for sixty days. Tim brought our stove and supplies, we were thrilled. Bless him. He delivers through rain and storms. He knows how anxious we all are out here. We were up very early to have breakfast in our trailer for the last time. After breakfast we went to work. Rob put the stove together and we set it on the tile. He hooked up the water tank from the bathroom into the stove.

We heard a loud noise and it was the backhoe coming up the road. We showed him the spot we wanted to put the well. He dug the hole and dropped in two cement rings. What a joy to have our own well! After using it for a few months we found we needed more water. That would come later. We were thankful for what we had now and learned to be careful with it. All the rain helped. Rob also hooked up the propane refrigerator. It was in the pantry. No more worrying about the precious food spoiling. The stove sitting on the tile, with the rock wall behind, was beautiful. Imagine one day a hot bath. The stove did have the reservoir in it, so I could dip hot water out to do the dishes. We talked about the well and Rob said he could put a ladder in the well and dig down deeper. We got the shovel, pick, a rope and bucket and put these by the well for our morning job.

Secluded Rendezvous

Rob put the ladder on the side of the well and lowered it. He dug the hole deeper. He filled the bucket and I pulled it up by the rope. I emptied the dirt into the wheelbarrow. When the wheelbarrow was full, I emptied it over by the garden to use there later. We did this for two days. Rob hit clay and sand and the water seeped in. We had built a water tank up the hill behind the house. We had bought a swimming pool for this storage. The water was pumped with our water pump into the pool. There was a water hose coming from the pool into a house line. The pressure was great. All was wonderful in the yard and house using this system. A person has no idea how precious water is until he's without it.

Chapter 9

John came by one morning, his boat would not start. We drove him back down to Marks Bay and brought back his battery to charge on our generator. After charging the battery we drove John back home. He gave us a five gallon container of gas. He was so grateful. So were we. Judith was away for a week. She asked us to watch her place. On our way home we stopped by her place. Rob was putting gas in her generator. He pulled the hose out of the tank and it was full of gas. The hose swung around and the gas flew in my face. We ran in Judith's house and thank God, she had water in her tank. We ran to the sink, Rob pumped the water pump. I put my face under it. I would have been in trouble had the tank been empty. We had an early dinner and dropped into bed. We had no trouble sleeping. Before John got there we had already done a days work. We worked in the yard and chopped wood.

We left early the next morning. We went in to Campbell River. We picked up our new phone, took back a chainsaw that didn't work, went to the store to buy what we needed, and made our trip back home. I couldn't wait to get home and try making bread in my new stove. When we got home, Rob built a fire in our new stove and I started the bread and yogurt. We had hot bread with our dinner and it was so delicious. It rose beautifully in the warming oven on top. The fire was just the right temperature. The stove was wonderful. The next day, we went fishing at Surge Narrows, in front of the store.

The owner of the store yelled to us, "Hey you two, come in for coffee." We tied up at the dock and watched the mail float plane take off—I love the little planes. I thought, someday I'll fly one of those.

We went inside the store and Mearne had the teapot on the top of the wood stove. There were four chairs around it. Jean was the other lady in the store. They lived in a nice house at the top of the hill. Mearne told us where her favorite places were to catch fish. We were not to tell a soul. And we never did.

While she was talking she said, "Caroline, you need a small Dingy with a little motor to come here and fish, but only on a good day." Robbie bought me the boat and a five horse motor. We towed the dingy home and tied it up. I later painted it, blue and white and gave it a name: "Going Dingy." Before I left California, I had sold one hundred and fifty photos. I have not done any shooting since, no time. I really was missing the dark room, framing, "Photo talk" in general. Mearne and I talked about it one day. She offered to put a few of my pictures up in the store, and she sold them. I was pleased.

The next morning we saw two people walking up the road and into the driveway. We went out to meet them. They introduced themselves as Ken and Marlene. They had just bought on the other side of the island. They were such nice people and we immediately became friends. Marlene and I had a lot in common: gardening, crafts, cooking, and other interests. We had tea and showed them around our place. What a fun time it can be to exchange ideas and plans with someone who is on the same path as you. They bought a house that sat on the water's edge in Campbell River. It was built in the forties, and belonged to the Thulin family. Ken and Marlene were having the barge named the "Totem #1" bring it to their property. It would go just far enough back from the beach to be safe from the sea.

Secluded Rendezvous

They had friends and family coming out to help them. It was wonderful. It was a lovely home, all complete with two bedrooms, a bathroom, kitchen, dining room and living room, and it was charming.

Chapter 10

Winter was here and Rob and I took a float plane into town. We had done this several times before. Quite different than the boat, car, ferries, not to mention the rain and cold. Having said that, I love the boat. This is just different and fun too.

It is too expensive to do all the time. We had some shopping to do. Marlene introduced us to "winter clothes" we hadn't thought to buy these. We just thought, it's winter so it gets cold. We would not be cold again. We couldn't wait to try them. The list of learning is long and hard. We left the airport, got a room at the hotel, and started to shop. We were able to purchase everything we needed. When we arrived at the Seaport to return home it was very foggy. I asked the pilot if it was safe to fly. He said, "Oh yes I've seen worse."

We got in the plane with all our paraphernalia. As soon as we took off, Rob puts his head against the window and fell asleep. I never slept. I love the planes and am too excited. Well, one could not see at all. The plane flew, and flew, and flew, and an hour later I woke up Rob.

I said, "Robbie, something is very wrong. Our, twelve minute, ride has been one hour, and we're still flying." "Oh, you must have read your watch wrong." Of course he is still asleep. He looks out the window and says, "Where are we? Hell, we are back in Campbell River." We looked out the window and could see the water, and we were about to land.

When we landed, I said to the Pilot, "Where were we?" He looked as white as a sheet and answered, "I don't know. I was lost." The people in the airport were happy to see us, everyone was waiting for some word. The owner said, "Sit and have a coffee and we'll try again in about an hour."

I said, "Call us a cab, and we'll see you in the morning if the weather is good." We went back to the hotel. The next morning, we took the first plane out to the Island. It was lovely flying over the water with the islands all around us, and no fog! Good to be home again.

Rob worked on the truck and I worked in the yard. We had been invited to Judy's for dinner. She now had two children, Ian age three, and Jenny age one. Ian used to walk to our house with a note pinned to his jacket at age two. We can't see each other's houses and this used to scare me. I was afraid of a bear or cougar, or his going the wrong way. But he would tap on our door. The note always said, Ian wants to play for an hour. We loved his company. Ian was a darling boy. We would play and have cookies and milk, and Rob would walk him home. One day Judy came crying, "I can't find Ian, have you seen him?" She had gone to several places and the whole island was looking for Ian. They found him floating in Marks Bay. One of the men brought him back to life. Rob and I ran in the opposite direction towards Ken and Marlene's to look for him. We knew every one else would be at the Marks Bay end of the island. By the time we found out about him, Ian was on his way to the hospital, in Campbell River. We were all sick. I told her never to send him out alone again. She said he had left on his own this time. I really think people out here try to teach their children to be tough and learn, too young. In my opinion, they grow up too fast as it is.

The island is growing. More people are moving in now. All excited, as we were, with dreams and plans. It

Secluded Rendezvous

is a new way of life and living. Sam lives up the road from John and Carol. He lives alone and has a real cute house. He built it himself. He has a truck and drives up and down the hill to his boat, and to visit. Sam is a very nice man. We went to see him one morning and our truck died in his driveway. He and Rob worked out in the rain and got it running. I was cozy in front of Sam's stove. It had a nice little fire. Sam had told me to go in the house and make myself some tea. I had tea ready for them when they came in. Sam is a very quiet man. He doesn't say much, but when he does it has a lot of meaning. He collects model cars and has a lovely garden with vegetables and flowers growing all together. It is lovely. He also has blackberries and strawberries. He owns an old wood boat, called the Sculpin. It has a place to sleep and a stove to cook on. There is a wood stove for heat. He fixed all this up himself. He is very talented. We stayed all afternoon and had a nice visit. He told us stories of living with the Indians for a while to learn how to carve. He is a beautiful carver. On our way home we stopped the truck and walked up into the mossy apple trees at Marks bay. It was a lovely sunny afternoon. There was a soft warm wind blowing, this was my favorite kind of weather. When that wind hits my face with the sun shining, I hold up my head and say, it's Randy on my face.

Rob went to town one day, to get a few supplies. I baked bread, made some tea and went outside with Pokey. We took a walk down the road and the tea cup was in my hand. It was a crisp morning and the sun was shining through the trees, it was so still and quiet. I could not hear a sound. Pokey sat next to my feet, as I sat on my log. Pokey looked up at me. I whispered, Pokey, I know each day we wake is a gift, with many things to discover. Beauty is here for us, and it's all waiting for us to uncover.

Carolyn Begg

I said, "Thank you God for this unbelievable peacefulness of beauty surrounding us." I sat on the ground next to Pokey, and hugged and rubbed him. I loved this dog. He was so good. We went back to the house and sat on the bench, next to the pond. The frogs were there. We watched them and they did the same to us. Now I know Pokey understands me when I speak to him. I began to talk to the frogs. They sat on the rock and stared at me. I thought you have a heart and a brain; you can talk to each other in your frog language. How do we know you can't understand us? I don't know! So, I told them how happy I was when I discovered they had moved into my pond, and I was happy they were a part of our family. I also told them they were darling and Pokey thought so too. I had finished my tea and went in for another one. I came back outside and thought, living here, I feel I am a part of the earth, sea, and sky. I am a part of something older and wiser than man. I get to be myself with thoughts and feelings all my own. I stand in the woods and stare at the green around me and say, "Lord, I can feel you here with me and I feel a burst of joy. I am thankful for your presence and I thank you for all the gifts you give me. I am thankful to you and your Father."

Marlene called and said Rob had driven by to see if Ken wanted to go to town. Ken was going to fish this afternoon, so he told Rob no thanks, but maybe next time. Ken said, "Marlene call Carolyn and see if she wants me to show her how to fish." Marlene said she didn't want to go, and if I would like to, go down to Mark's Bay and meet him there. He has all the gear. So, I did.

We went close to the island and put out our lines. We were sitting there with our lines dragging behind us, and talking about our plans with our homes, when

Secluded Rendezvous

I got a fish on my line. It was a big one! Half the way through I said, "Ken, please take my line for a while, this is a heavy one."

Ken being a very good fisherman said, "He's your baby Carolyn." He made me bring it in. It was about 25 pounds. Ken was so proud I did this—and I was dead tired. I told him just wait until I get the next one. He laughed. In just a short time, I did. I was so happy, and laughing so much I thought I wouldn't be able to reel it in. Ken did take over for me and brought it in. The best part of the story is—Ken never caught one, and I never let him forget it. Ken told me, "Every time I go fishing, Marlene says, 'Take Carolyn with you, and laughs.'" I have a lot of fun with these two.

Chapter 11

One month has passed and another has begun. We are driving in our 16 foot skiff to Campbell River. We are going around Quadra island. We will land on the dock in Campbell River. The water can be wild there so we don't do it too often. Our Correct Craft is being fixed. When we arrived it still wasn't ready as promised. We got a motel. A huge storm came in and we had to stay another night. In the morning were towed from the garage to the waters inlet where the boat was put in. We unloaded our truck into the big boat. It was filled with full gas cans, groceries, plants, paint, tools, fruit trees, and other goods. There was hardly room for us. I sat over our motor with my lap and arms full. We had left the small boat in Heriot Bay. We tied it to the big boat. We would tow it home. We pulled out from the dock, got out in the bay and the engine started to smoke. Remember, I was sitting on it.

Rob being his usual self said, "Don't worry, we'll make it home."

I said, "Not me... take me back to the dock." We had a few words, and went back to the dock. The man who towed us was still there. Other people were on the dock watching and worrying about the boat with smoke coming out of it. We hired a kid to drive us home. The boat went back to the shop. The boat was brought back to the island. We tied it up, dropped an anchor, and one night, a storm filled it. This boat was taken back to town and sold, as is. It was really too big for us to care for. We were too busy on land. The boat was better off and we were happy. Our other boats worked just fine for running in to town or fishing.

We decided to go to Victoria one night. We left in the morning for Heriot bay, took our car across to Campbell River, and drove two hours to Nanaimo. There we took the ferry to Victoria. Nanaimo is a city in south east Vancouver Island. It is a small city. The population is about 49,000. We landed in Victoria. This is the Capitol of British Columbia. The population is about 64,000. There are many tourist there, as it's a lovely place to visit. The beaches, parks, shops, and buildings are grand. There are flowers everywhere. Hanging baskets are on the lampposts all along the street. Horse drawn carriages take you sightseeing. It's breathtaking. We stayed in the Empress Hotel. It was built in the early 1900's. We spent the day playing "tourists." We had dinner in our room. The next day we bought a water pump and another phone with 10 channels. The phone lines are very busy "out there." We had "high tea" and left for the Ferry. In the town of Courtney we bought two trays for the laundry/entry room. We bought a wringer washer, Radial Alarm saw, and water line. We bought a refrigerator and a table saw. We were told to have two refrigerators and use one for a freezer. We stayed in Campbell River overnight. We saw Tim, and arranged for him to call us when he could deliver everything. We did our shopping. I went to my doctor who said I wasn't built like a man and I was working too hard. Of course I didn't think so, but I took it slow for a couple of days. When we got to our dock and up to the truck it wouldn't start. I guess it was "working too hard."

Rob said, "You sit here and I'll go up and get the wheelbarrow." I was happy. It's hard enough walking up that hill, without the wheelbarrow... poor Rob. When he returned, Rob looked tired. We started up the hill. The wheelbarrow was full. I took a few things out of the wheelbarrow, and carried them, it helped.

Secluded Rendezvous

We stopped along the road and drank some juice. We got up the hill and turned left to the flat road toward home. When we came to Judy's I yelled, "Oh no." Rob let go of the wheelbarrow and ran to me saying, "What's wrong?"

I cried, "Look at the yard." We opened the gate and went in. The deer had been there and trampled almost the whole garden. The fruits and vegetables were pulled out. It was a terrible sight. We found a hole in the fence behind the house. Judith had gone to town and decided to stay overnight. We went home and put our supplies away. Rob built a fire. We had tea and decided we were too tired to eat. We ate a peanut butter sandwich and went to bed.

Thanksgiving was coming up. Thanksgiving in Canada is in October. We thought we would ask Peter and Edde to dinner. We walked down to their place early the next day. We asked them and they were very excited. We went home and made plans. I cleared the bracken under some trees, moved some rocks and raked the area smooth. Rob built a long table with two long benches. It looked great! The next day he said he would take the boat in and get a turkey. I made him a list of what we would need. I stayed home and baked bread and pies. What fun, baking with my new stove. My first big dinner. The next day was Thanksgiving. The company walked up the road carrying food and wine. I stood on the deck and thought just like it's written in the old days. How lovely! Peter and Edde brought their son, Michael and Dieter were there too. We all had a wonderful day. It was all about their plans for building. They were going to have a pre-fab home. It would have a big kitchen, a living room, two bedrooms, and a bath. The house would all be brought in by helicopter. Peter was having five men from Germany come this summer. They would help him put the house up. They were told that it would take three days to do it. We had a beautiful day. Peter brought a very nice bottle of wine. Edde brought baked goods. Our guests were nice, fun people. We all sang and laughed and had a great time. They said we must come to their house when it is finished. Edde laughed and said, "No we will eat in the old house. We will not wait." We told them we would love it, just so were together.

Chapter 12

Rob and I played for a few days and talked about Christmas. Everyone would be away with their families. Peter and Edde would be with Dieter and the rest of the family in Vancouver. Rob and I tried to be enthusiastic about it. It was our first Christmas away from our families. Rob went to the post office and came home calling my name. I came on the front porch he was smiling and carrying boxes and lots of cards. We took them in by the stove. The boxes were from our families. They all sent presents. His dad sent some that had writing on them. The writing said open before Christmas, and we did. There was a whole canned chicken, chocolates, instant potatoes, cranberries, candy canes, a music box, and tree ornaments. Randy sent warm gloves. We needed these as they were for cold Arctic weather. We received gifts from my Mom, Rob's Mom, and my sisters. We would open these tomorrow. Rob stood up and announced, " I'm going out and cutting a tree. Decide where you want it while I'm gone." He came home with a beautiful tree as tall as the ceiling. We decorated it and put all the presents under it. We were like little kids on Christmas morning. We got our gifts from each other and put them under the tree too. I cried.

I went out on the porch and said, "Son, mom will be all right, thank you God." We strung popcorn that night and put it on the tree. I said how lovely everything is, all we need is for it to snow. Rob said, "It's cold enough." We woke up the next morning and it had snowed during the night. What a breathtaking sight! The trees were all dusted with snow. It looked like a photograph. We had our radio on all day the program had Christmas music on all day. At night there were Christmas stories. I told Rob the people in Vancouver did not forget us out here. We called every one in our family and some friends too. How very fun. I must confess, I had sudden tears during the day. I never knew if it was from joy or sadness, but it didn't matter. We were all happy and well in our family. Maybe it was for those who could not be. We later took the tree outside and many birds came and ate the popcorn. It was great to see!

A few days later we called a plane to pick us up and take us into town. We stayed a couple of days and bought lots of much needed supplies. Tim was bringing out our load from our last trip. The flight home was spectacular. The water was deep blue and the trees were dark green with sifted snow all over them. It looked quiet and still. We knew we could not make it up the road from Marks Bay so we landed at Harold's dock. The pilot helped us unload our boxes and groceries. There was deep snow on the dock and everywhere. There was an ice wind blowing. The pilot opened his door and turned to look at me and said, "I have to take off now will you be alright?" I assured him we would be. He climbed in the plane and took off. I watched him and he tipped his wings back and forth waving at us. We watched him until he was a tiny snowflake in the sky. I had a lonely sunken feeling inside. It was so still all around us. I wondered if we could carry all this up that snow covered road? We wanted to hurry because of the cold wind. But, "hurry" was not to be permitted. We carried what we could and decided to walk slow and steady. We would return to the dock after a rest at the top, and continue until we had it all up. Then we will take it home in shifts. We were taking the first load to our home when I started getting pains in my jaw and down my arms. Rob told me he would take it home and for me to stay there and rest. When he returned, Judy was with him. She had heard the plane at the dock and knew it was us. They carried our things and we all went to the house. When we got there, she had a fire going in our stove and had brought bread and cookies. Ian was a darling, he brought me a cookie "Cause Carolyn didn't feel good."

Between the road and garden, we had started a burn pile back when we were living in the trailer. This pile was where we put our boxes, bracken, branches, and anything burnable. The pile was about ten feet high and twenty feet around. On these Islands, one could burn in the winter only. You can put old tires in the pile for weight and a fast hot fire. For days you stayed with this fire-day and night. We did this together. You are constantly raking the edges and pitch forking into the pile. It is a scary thing to work on with all the trees around you and the other islands. You also keep a hose attached to the pump that is going into the well. The fire pile is very hot, sitting on the ground with tree roots under the dirt. We were constantly looking for smoke coming up from the dirt where we raked. Each night we watered around the circle. This was a very nerve racking job. We were quite happy when it was over.

We decided one night, that it was time to have a real outhouse. In the morning, bright and early, we went out on the path from the bedroom deck and began. I cleared a space and Rob dug a 4x4 by 4ft. deep hole. He built the walls around it. This room would be eight feet long and eight feet wide. The walls were six feet high. He had a front door and a slanted roof to the back. He dug a trench across the back with rock in it. Inside to the back wall was a long box from one side to the other. It

Carolyn Begg

had two holes cut out on the board on top. We had a regular toilet seat over one hole, and a cover on top of the other. Every few months we would switch the seat. He built a magazine rack on one side of the wall. There was a window on the other side three feet long and two feet wide. We covered it with screen. The roof hung out over the door to keep the rain out. Rob put a light blue, wall to wall, carpet on the floor. I wallpapered the walls with flowered light blue and white paper. I had scented candles and a basket filled with cedar sawdust. When you used the toilet, you put a handful of sawdust down the hole, thus collecting moisture and odors. It worked very well. We had an outhouse with no odors, a rare thing. I hung pictures, and had a potpourri basket. We had an outhouse so fine, it was the talk of the island. People would come to our place and say, "We came to see Carolyn's outhouse." We had a lot of laughs over it.

Ken and Marlene walked to our house one day. Marlene made the best beer you ever tasted. Hers was the only beer I could drink. Ken carried some to us. As we sat drinking the beer Ken said, "I killed a deer last night with one shot."

I said, "Why did you do that?"

He laughed and said, "For food."

Marlene said, "We do it all the time; I can it for winter."

After hearing their story about too many deer on the Island and that it is too far to go to town to buy meat. Then he said, "What's the difference if I kill a deer, or go to town and by a piece of beef?" I thought this is okay, it makes sense.

Two days later Rob got out his rifle and said maybe we would do this too.

I said, "Robbie do you think you can kill a deer?"

Secluded Rendezvous

"Oh sure" said Rob.

So off he goes and I yell "Be sure you aim at his heart or between the eyes."

"Of course." says Rob.

"Don't pull the trigger if you can't hit your target with one shot," I yell. So, I am standing on the front deck waiting and I hear a single shot. I wait and wait—pretty soon here comes Robbie up the driveway, with his head down. I holler, "Did you kill him?"

Rob said, "I think so."

I said, "Where is she, maybe she's hurt."

She was across the road. Rob and I went walking over to her, and she was dead. We looked at each other with tears in our eyes. I ran home and got the hunting book and read it to Rob. He carried the deer home. We did everything by the book.

When we saw all the meat wrapped on the table, we were excited about getting our freezer full. We put it all away, cleaned up and looked at each other. Rob came across the room, put his arms around me and I cried, "Oh, Rob, the poor thing, let's not do this again."

"Never," was his reply. Pokey didn't go along with any of this either. He sat by the door of the shed the whole time it hung. We never did it again. I do have to admit it was the best meat I have ever eaten, and we did eat it. It was not going to die for nothing. I was happy when it was gone, out of our freezer. We ate fish and chicken. When we caught a fish, I took the stick and hit the fish between the eyes with one hit. I never let them flop on the bottom of the boat for hours. I learned to kill a chicken at the age of five. One flip of the wrist and it was done. Nothing is easy to kill, but if you have do it, don't make them suffer. I was proud of Robbie doing his one shot. I never used my hunting license again. Rob and I lived in the house another year.

The kitchen cabinets were in place. The water system worked very well. The wringer washer and tubs were heaven sent. The bathtub was grand. The garden and fruit trees did very well. The work was all done, and we were happy with the results. But now what? Rob came to me one night, and said, "Carolyn, the money is getting low. We need to figure out a way to make money. Everything here, in Canada, is three times more than I planned." I had no idea. I had let Rob handle all the money. I had done it with my first marriage, and was happy to let it go. Our original plan was to be here in Canada during the summer, and work in California during the winter. This plan never happened. When we were in bed that night, we talked for hours. Rob fell asleep and I got an idea.

Chapter 13

He woke up in the morning to, "Rob, let's build a Fishing Lodge, down on our beach."

"We can't build a Fishing Lodge, Carolyn." said Rob.

"Why not? We built this house. We just need to build one bigger," was my reply. Then Rob thought about it, and away we went. The excitement grew by the hour. We sat around the house for a couple of days making plans. How big should we build? How many people could we handle at one time? Maybe we should have a place for seniors, or Weddings. How about a home for boys? Being way out here where the fish live was the answer. We sat down with a big piece of paper. We thought to have five bedrooms. There would be a long hallway with two bedrooms on each side with a bathroom in the middle of each one. One side would be a bathroom for men, and the other side for ladies. We would build a living room, a dining room, a kitchen with a pantry, and another bedroom for the cook. We would have a window seat, library, and a huge fireplace in the living-room. There would be a laundry room. The upstairs would be where we would live. We would have a bathroom with a tub and a shower, a dressing table, a window and a skylight. We were planning as fast as our heads could think.

The next morning we got up at dawn and took the boat to the other side of the Island. We stopped the boat near shore. We were looking at the ten acres of ours, and picturing where we wanted to put the lodge. We tied the boat to a rock, and walked up into the trees. The rocks at the water's edge were about twelve feet high. This would make a wonderful break for the waves at high tide or in a storm. Walking up from the shore the land sloped upward to the end of our ten acres, meeting the logging road at the top. Our house was just off to the right. The house would be up a 250 foot bluff from the lodge. This would mean a trip up and down twice a day while building the lodge. The island was solid rock with trees and ferns. We did not have to worry about the hill or house coming down on the lodge. It looked like the perfect building site for a lodge.

Louie was working at tree cutting, so we hired someone else to come and look at our site. We didn't feel we could cut the trees we needed in our area without making a pile—one tree on top of another. We left it to the people who were experts in tree cutting. When the man finished cutting, the trees were on top and across each other, not something Louie would have done. We took our chain saws and started cutting the side branches off. We drug them out and pulled them to a new burn pile. We kept this process up for days. Every night we walked up the bluff to our house. It was a long climb. I cooked dinner, we bathed and fell into bed to get up at first sunlight. A light breakfast and down the hill to the building sight again. We separated and rolled the trees into a space, this was a place next to the Alaskan Mill. These trees were heavy. We used a peavey to do this rolling. We left the trees for drying and went on with burning and clearing.

We discovered a very large tree stump where the kitchen was to go. Rob said, "This is in the way of our building." I said I would dig it out. "You can't do that, it will take you a year, it's too big." Rob laughed.

One week later, Ken and Marlene stopped by. They walked by Rob, and Ken said, "Carolyn, what in

Secluded Rendezvous

the h--- are you doing with that pick?" I answered, "I'm digging out this tree."

Ken shouted, "Are you nuts, that will take you the rest of your life." We all sat for a while and talked. They went on their way and we went back to work. It was always good to see them.

I used my shovel, pick, and saw, and I dug the tree out. I worked on it for two weeks, and it finally came out. It was four feet across and I thought I'd never see the end of the roots. I cut pulled, dug, and prayed.

When Ken and Marlene came two weeks later they couldn't believe it. Neither could we. Ken told Rob he knew where there was an excavator on Read Island, and he used to drive one. He and Rob took a drive in the boat. I could hardly believe that maybe a machine would be moving rocks and stumps for us. A great idea!

I took a break down on the beach while I waited for them to return. I sat looking at the waves and listened to the song of the birds and the melody of the sea. It was wonderful to be sitting on the edge of the water. I saw my reflection in the water as I stared at my face, I asked, "Where's the girl who sat in front of her mirror putting on her makeup?" My wants and needs have changed. I have Robbie here and I love him. But I miss you Randy... my son. I miss our long talks, our laughs over silly things that made sense or not. I miss walking with you, and holding your hand, and then your holding my finger. My heart is saying, "Oh." My most treasured memories.

Ken and Rob came in the boat and splashed me with waves. Laughing. They said Rob would get a small barge to bring the excavator to us. Ken would start in a few weeks. Ken went home.

Rob and I continued to work with the trees. We cleaned the area and drug and rolled the trees off to the side of the saw. We talked about bringing the tent down from the house. Rob set it up one day. We put a tarp on the floor, wool blankets, and our opened sleeping bags, sheets, and more blankets. It was very warm once you got into bed. We soon got a propane heater. That was great, we thought. The winters in Canada were very cold, the kind of cold that goes deep into your bones. The air was so fresh it was unbelievable. I used to say taking a deep breath was like drinking a glass of ice water.

I said to Rob, "When the foam blocks come, let's put them all around the tent and make walls." He did that and it kept the cold and wind out, it was great. We lived in the tent about three months. It was good to live next to our work area. Not knowing propane heaters gave out moisture, we later found our bottom layer of wool blankets to be damp. I had a deep cough for about a month.

I woke up one morning with a fever of 104. We went to the doctor in Campbell River. He was not happy with us. He said, "She has a lung infection."

I turned to Rob and I said, "I guess I should go up to the house for a couple of days to sleep."

"Where are you sleeping now?" the doctor wanted to know. The tent was not something he wanted to hear. He gave me some pills and we returned home. We went up the house, Robbie built a nice fire, and I went to bed. I slept for hours. Rob stayed home the rest of the day. It had been cold and raining and we both needed a rest.

Two days later Rob came home from the lodge sight and said, "Carolyn, the tent is gone. Ken fell a tree on it with the backhoe."

I replied, "Maybe it's for the best, but we'll miss it." When we were in the boat coming home from the

Secluded Rendezvous

Doctor, I had said I would stay in bed in the tent until my fever broke. I didn't feel like climbing the bluff. Rob said he would help me up, I needed to be where it was warm. We were sure glad he took me to the house. A tree landing on me would not be a good thing. Imagine how poor Ken would feel? Jesus was with us again! We were told later that Propane heaters give off moisture and fumes, never keep them enclosed.

One morning Ken came and worked on moving rocks and stumps. It was wonderful to be able to say, "Just put that rock over these Ken." We would yell above the motor, and he would give us a look from under his hat, a nice man with a good since of humor. Good to work with. You know that stump I dug out? Well, there were two of them. He dug and pushed it around for hours. I told him I could have had it out in half the time. When it came out, I asked him what took him so long?

It really was unbelievable to watch him work that machine. I had discovered a place where there were huge walls of rock on two sides and dirt in the middle. It was wide, but not wide enough for the machine to get into. That area was where our living room was to be. I thought, I'll dig that out and see what's under there. Maybe it will be okay. I showed the men and here came, "You can't do that Carolyn."

I got my pick, shovel, and rake, and started working on the dirt. It wasn't so hard because the dirt could be pulled out the bottom, and into the water. After moving dirt for a week, I told Rob to help me run a hose from a pond up the hill to down here. I wanted the water to help me, and it did. I cleaned off the sides of the rocks, and Rob was so happy he said, "This will make a wonderful cellar. We will have about 14 feet of floor space, and it will be about 8 feet high. The rocks on each side will keep it cool." We left it, and returned to our work.

Carolyn Begg

Ken was working on a big job digging a large hole for a septic tank. We were so thankful for him. Imagine us digging it out with a shovel. He worked a few weeks more and was finished with the work. We were so thankful for all that he did. He and the machine went home. We would miss them both.

Chapter 14

Rob and I cut a lot of Cedar trees with an Alaskan Mill. He ran the motor and I was on the other end of the blade. I was guiding the blade with a handle. The ground was clear, where the motor was. My end had dirt and rocks to climb over. There was a lot of moving your arm up and down. My arm was hurting for a week. I finally told Rob I had to go to the doctor. He said I had tendonitis and could not use the arm until I had no more pain. We stayed in a motel, had a nice dinner, watched T.V. and returned home in the morning. I had an air pillow on my arm. It didn't hurt, so I kept working with the saw. Two weeks later, the other arm had problems. The doctor said I was not to use my arms. We went back home. Ken and Marlene came by. I told them my story and Ken said to trade ends with the saw. It's a lot easier to run the motor than the other end. We did this and he was so right. I healed somewhat. We finished what we were working on and I was happy.

Robbie built a small structure and covered it with a tarp leaving the front open. It had two shelves across the back for a burner to cook on, and our cups, tea, and chocolate. We would bring our food down each day. It was nice to have a kitchen. We ate bananas, apples, and noodles in the Styrofoam cups that we filled with hot water. It was nice to have hot soup on a rainy day. We were now working on trees again. If you can imagine a huge tree falling, cutting its branches off, cutting these up to size, putting them on the burn pile, then pulling the tree to the saw and taking a slice off the top. This was one tree and we had many to do.

We were in our forties. Rob is five years younger than I. He likes to tease me about it. It was raining and raining. We had good Halley Hansen rain gear, gloves, and nice leather boots with lining inside. We were dry and very comfortable. After work we would walk up the hill to our house. I was usually sitting each trip. My heart didn't like the climb after a day's work. So I sat while Rob went ahead to start the fire. We knew I would be fine in five minutes. I was told by my Cardiologists to sit when I had pain in my arms and jaw. It only happened when I walked up the hill. I could pick up a tree by myself and be just fine. Maybe I just wanted to sit quietly in the ferns, and have a warm house to go too.

After dinner we had a nice hot bath, and fell into bed. In the morning we had breakfast and went down the bluff to work on trees. The trees we were slicing were Cedar. They would be the underpinnings, bottom layer, set on posts. We bought a cement mixer to make the pads to set the wood on. It was a real blessing, from mixing cement in the wheelbarrow. We really learned to appreciate the things we took for granted in the city. We used our Honda generator to run the mixer. Things were getting easier. We now had a 70 x200 foot clearing for the lodge and the trees were done.

We needed to make a trip to Vancouver. After deciding to chance the weather, we got in our boat and headed for Heriot bay. Rob assured me it would be okay. When we got out a few miles from the island the swells were high. Rob cut across the water and got in a trough and yelled, "Carolyn, sit in the bottom of the boat." I did this. I could not see land. The water was walls on both sides of us, and we were riding in the middle, cutting across the Sea heading for the dock at Heriot Bay. With spray hitting my face, I said, "Jesus, please get us out of this and I'll never come out here again when I

Secluded Rendezvous

know I shouldn't." I never did it again—knowing when we left, we were taking a chance. We got to the dock. I said, "Thank you, Lord." We got our car and drove on to the Ferry. It was a rock and roll ride. The cars were all washed when we reached the other side. The people stayed in their cars. They were afraid of being washed off the deck.

We drove to Nanaimo, on to that Ferry and on to Vancouver. We arrived at mom and Haakon. They were happy to see us. They knew the weather was bad. We had dinner and sat and talked. It was always a pleasant thing to do with them. In the morning Rob called a big lumber mill and put in an order for lumber. We decided not to mill all the lumber for the Lodge. We also had to hire a barge to bring it. He ordered sinks, toilets, showers, siding, pressure treated lumber, roofing, and basic supplies. This barge was an old steamship. It was about 70 feet long, a great ship. Rob and I, as usual, enjoyed our stay with mom and Haakon. We needed to get back home. The weather was good now and we had to run for it while we could. We were all sad to leave so soon.

We had a nice trip home. The water was smooth. We had done a bit of small shopping in Campbell River. People "out there" are always out of fresh fruit and vegetables. We had dinner and then drove home. We were on our Island before dark, and soon in our bed fast asleep. In the morning we had to figure out where the best place for the barge to come in to shore and unload our supplies would be. We couldn't move it once it was placed. We found the best spot for them, and for us. We had trees to cut for floor joist, across the front of the basement. We stacked these off to one side. We cleaned the area and were ready for the barge to come.

We stayed in our little house all day and did nothing! Unbelievable! We took a walk, sat on the deck, and drank tea sitting in our rocking chairs. We talked and made plans. The barge called at dinner time. They would be arriving sometime the next day. I don't need to tell you, we talked all night about where and how we would stack the pile. It was a large one. We woke up to the sound of the hum of the barge. When we arrived at the beach it was not in sight. The noise carries a long way. About an hour later it came. We were so excited. The barge landed and it had a crew with it. We all unloaded and stacked the supplies in place. As things were being stacked I was saying, "Oh good, we ordered this oh boy, boxes of this and that." It was hard to remember if we ordered everything we would need—even though we had gone over the list many times before ordering. The barge left and we stood in front of the pile. The cost of delivery was a few thousand dollars. They did come from Vancouver and had to pay the crew and gas. We were thankful for the fast service and the crew.

In the afternoon Rob took the boat over to Ken's to see if he knew of someone who wanted a few days work. He told Rob two kids were camped next door to them, and were going to build a home. Maybe they needed some extra money for supplies. Rob found Mike and Charlene on the property. Mike expressed he would be happy for a few days work. I stood looking at our pile. I thought, this is a huge pile. Like a big puzzle, and we have to put it together and create a lodge. There were boxes of nails, screws, roofing, shingles, rolls of tar paper, sand paper, paint, sealer, showers, sinks, toilets, pressed wood, flooring, stacks of 2x4's. The list was long. I thought this was my idea. Rob and you can and will do it, keep your excitement up, gather ideas in your mind and keep your faith and common sense together.

Rob returned and told me he hired Mike. Harold sold his place to them. Rob was working on the basement putting pillars in to hold the floor that would be the living room floor. He dug holes, and put in the cement. The form for the cement was a box one foot high. He placed a five- inch slab on each one. We peeled these and brushed them with the preservatives. The pillars were every six feet. These pillars were to hold trees twenty feet long.

Mike would be there in a couple of days. Mike and Rob would put these trees on. When Mike arrived he and Robbie began. It was a heavy job. I got to rake, carry tools, and make tea for them. It felt good, easy job. The job took a few days and Mike was back home. He was happy with the extra pay, but anxious to get back to his place. The next day his wife came to meet me. Her name was Charlene. They were just starting to clear their land. Being there longer I got to give her a few tips. I don't believe she needed them. She seemed very self-sufficient. I told her the men out here will ask you to pick up the end of something (like a tree or a framed wall). Knowing it takes two to do these tasks—you do it! Neither one wondering if it's too heavy… you just do it. It's amazing how the strength comes to you. After doing these things you become as strong as the men are. When you're young you don't realize what you are doing to your body.

Robbie and I began to work on the rest of the building. This was an area being 70 feet by 40 feet. We were putting down cement pads, posts on top (painted with preservatives) and a square of roofing on top. We used 8x8x13 beams and 2x8 boards. These were 16 inches apart. We had a pile of beams that we had milled. On the top, we used sheets of flooring. These were tongue and groove. Under this flooring and in between the 2x8 I cut rolls of Styrofoam and placed them in between. It took days to do this part of the floor. I arrived to finish my job one morning. I was counting how many more I had to cut down at the end of the house. I was looking across, as I walked, when I stepped on a piece of Styrofoam and I fell through. I called Rob. I told him I was dizzy. He came over and pulled me up. He said, "I can't believe you did that, Carolyn, You have been working on this for days. You know you can't walk on this foam." He turned to walk away, stepped on the foam and fell through. The 2x8 board scraped his leg all the way up to his hip. He had a bulge the size of a grapefruit on the side of his hip. I told him it was his punishment for yelling at me. We took him to the doctor in Campbell River. The doctor said it was the cellulite being pushed up into a ball. Rob was a thin man. The doctor said it would absorb back into the skin. It took over a year to do this. When the flooring was down, we raked and burned to clean up. We did this every time we finished a job. It made the next job so much cleaner. We also did this to keep our fires small. I had nailed the whole floor down. There was a nail every six inches. These nails were galvanized twist nails. The framing nails are longer, but straight and much easier to sink in, but not as strong. We began to build the walls. Doing the walls was fun. You build them on the floor. When they stand up there is the whole wall done! I was in a hurry to start the next one. We were actually very fast at doing the walls. We had practice in the house on top of the bluff. The hard part was carrying the 2x4s. Rob did this while I was nailing the floors.

We had a rainstorm one day and after working, in the rain, we did not want to walk up the bluff. We were too tired and the bluff could be dangerous as wet as it was. It would be dark before we got home. We carried two bathtubs into a bedroom. We put tarps on top of them and slept in them. The water came in mine and it

Secluded Rendezvous

was freezing. Very early, in the morning, we walked up the bluff to the house. We had a hot bath. The Oatmeal and tea were a great treat. It was still coming down in buckets. We stayed home all day. It was wonderful to be inside with the wood stove reading our books. When we were home like that I would cook dinners and put them in the freezer for later. It was fun and easier to do them ahead. We were ready for work the next day. We had to take our time getting down the bluff everything was very wet. We got to the building and it was cold. I said, to myself, "Don't think about it-pick up your hammer and you'll soon be warm." It wasn't long after when I was taking my rain jacket off. We had built out the window frame. Robbie built a window seat, it's lovely. After finishing that area, we heard a boat. It was Ken and Marlene. We sat on the new floor and I made tea. We still had our kitchen outside. Ken and Marlene had friends over on the weekend from Campbell River; they partied and had a fun time. Ken was holding his head and said, "Next time they get coffee." We had a picture of that and laughed. It would never happen. When there are friends over for the weekend it is party time. We had a nice visit and they went home.

We finished the walls in the living room. They were 20x40. The dining room was 20x18. This wall job took a week to do. After work we would walk the bluff to home, take a hot bath, have dinner, clean the kitchen, and up to bed. That was our week. We had done five bedrooms, three bathrooms, a long hallway (six feet wide) a laundry room, and a utility room on the end of the hallway to the right side. On the left side there would be a stairway going up to our living place. It was Friday. We put down our tools in the afternoon and walked home. We said we were taking the weekend off from the lodge. There were things to do in our home and yard too. We talked on the phone to the family, it was so fun to hear all the stories, and tell them all our news. They can't believe all that we have done. I miss them all. We were in bed early, and talked about tomorrow. The next day, while having tea on the deck, I suggested we go fishing for two hours. I wanted to can some salmon, we only had two jars left. We were out about an hour, caught two beauties and returned home. We each got our fish in the first fifteen minutes. We had a lot of things to be done in the two days we were off and we needed to get started. It was so tempting to stay on the water. The water was like glass and not a boat around. It was so still and beautiful. The sun was climbing up into the sky warming the Islands. We returned home. We cleaned the fish, cut it in one and a half inch slices, and put it in the refrigerator. I would can it after dinner. Rob went out to cut firewood and I filled the trays and washing machine with water to wash the clothes. I never minded doing the wash with this wringer washer; it's fun, actually. I like hanging them on the pulley line to dry in the breeze and sun. They are so fresh when dry. I finished the wash. I made lunch and helped Rob carry wood. It's so nice to see the shed full. I went in and canned the fish. Our pantry was full and it is a blessing to see.

The next day was Sunday and it was a nice day. I cleaned the house while Rob cleaned the woodshed. We wanted to go to Ken and Marlene's for a visit. I told Rob I would go ahead and wait for him just off the road. He was bringing in some wood. He soon found me sitting on a dead tree among the ferns, relaxed in warmth and timelessness. I said, "Sit, and listen to the complete silence. Jesus was here with me." Rob laughed. He doesn't believe in Jesus being God's son. It's hard for me to understand this, but to each his own.

Secluded Rendezvous

We got up and continued on our way. Ken and Marlene were out in their yard. Marlene was pulling weeds, and Ken was staking wood. Marlene yelled, "Oh boy, it's time to quit." I told her we could both pull weeds and she would be done. She told me she was so tired, she was glad we came. We went in and she made tea. She yelled to the guys to come in for tea. Ken told her, "Rob and I want a beer." We all enjoyed our drinks and had a nice afternoon talking. Ken and Marlene told us of their life in Campbell River. They lived there and raised their family. They were originally from the Eastern prairies, Manitoba. We heard many stories of how they met and what they did in the good old days. It was so fun to hear. Rob and I returned home, and asked them to stop by anytime. They said for us to do the same. That's what you did on the Island.

Chapter 15

Sam had a red flag out on his front porch. When it was there you didn't knock. While Rob and I were walking home through the forest of trees, the sun had a long day and began shimmering in and out of the leaves on the trees. We were both quiet till we reached our home. Rob lit the stove and I began dinner. Once we were settled, we began talking at once about Ken and Marlene and our lovely visit. Strange how you meet people from all over the world and you all wind up living on an Island to live the same sort of adventure, and yet our wants and needs are so different.

It was a bright sunny day and it was Monday, we were ready for work. It had been nice to have the weekend off. We would be putting the floor on upstairs (which would be the ceiling on the downstairs) it would be fun to see the downstairs rooms covered. What a difference it was going to make. Two jobs will be done at once. The boards were heavy to carry coming up the ladder, but it went pretty fast with the two of us. When the floor was done we build a huge triangle. It was forty feet long and twelve feet high. This was to go on one end of the building. We built it on the new floor and nailed sheeting on because we didn't have a ladder tall enough to nail it once it was up. It had to be nailed on the outside. We went to lift it up… it was so heavy. We got it halfway up and set it down again. We were afraid of dropping it over the side. While we stood looking at it wondering what to do, a boat came up to our dock. It was Peter and three men. Peter called out our name. We answered, "We are up here, use the ladder and come up." They came up and Peter introduced us, they had just arrived from Germany, and were anxious to see our place. They looked down at the triangle and asked Peter to ask us what we were doing with it. We laughed and said were not sure. These men could not speak English. Peter told them where it was going, and that it was too heavy for us. They went to the triangle, ho, ho, ho, and up it went. Rob and I picked up our hammers, and nailed it up. We were so thankful. Peter laughed and asked what we would do if they had not come along. Rob said, "Cut it in half, and get it up that way." They were happy to meet us and asked if they could walk to our house. We told Peter any day, and let them go in. I said there were cookies in the pantry. Peter turned to us and said, "Please come to dinner at our place tomorrow." We said we would.

The next afternoon we left work early and went home to get ready for dinner. We walked to Peter and Edde's. It was a nice walk down the logging road. They were sitting under the trees. They were very nice people and fun to be with. We went for a walk around their property and talked about their building plans. They were all excited to tell us about their plans. Edde did not speak English, but she understood some. She told Peter to tell me she was going home, to Germany, to go to school to learn English... and she did. Their plans sounded like they would have a real nice homestead. There was an old apple orchard there with lots of trees. There was also an old garden site just waiting for someone to come to it and rebuild a garden. They had a wonderful place for a dock, and a nice beach. There was an old house and a small building off in the woods that someone had lived in once. The property was 20 acres and had lots of possibilities. The dinner was superb. They were both excellent cooks. They each made us delicious German

Secluded Rendezvous

dishes. They had candles and wine, it was all lovely. I asked where the men were that helped us with our wall, we didn't properly thank them as they had hurried off. Peter said they had returned to Germany and would be back next summer. Peter and Edde's house would be flown in by a helicopter. It was to be a prefab building. It would take three days to put it together. Imagine that! I am happy for them. They work hard in the winter in Germany. They have a business there and are training their son to take over the business. They are looking forward to retirement. Edde served the cake we brought, with coffee, and after dinner drink. Rob and I thanked them, and walked to our house.

We were carrying flashlights and saw three deer on our way. It was a beautiful night. There were many stars forcing light in all directions. It was as though they were saying, look up at our brilliances we will shoot and flicker for you. It was a grand sight for us to be a part of. Some of the stars were dusty, too far to be clear. I felt as though I would burst by the time we arrived home. We were at the lodge early and we were going to build the rest of the walls upstairs. This was to be in a big room where we would be living. Robbie was going to build a platform for the tub to sit on. There would be two steps up to the tub. There would be a window along the side of the tub that would look out to the yard and water. From here you would see the other Islands and boats going past. I loved doing this with my candles burning and all my bubbles around me. A glass of wine would be nice too. It was to be a heartfelt delight! In our living room was to be a wood burning stove and a small wall for wood. Rob wanted to build three dormers in it. One was for an office for Rob. One would have a window seat, looking out to the water. It would be a sitting place for reading, writing etc. There would be shelves for a phone and books. Another dormer was for a guest bedroom. Dormers are wonderful. I was glad he had this idea. They give you so much more space and I feel there are little private spaces for you to sit. While we were building the upstairs walls, Peter and Edde drove their boat by blowing the horn and waving goodbye. They were taking the boat in for storage. They would return to Germany and be back in the summer. I'll miss them.

We worked a couple of weeks and Rob said, "A couple of guys will be here in a few days to help." He and two men would put the roof on. I was happy. We finished the upstairs. Rob built the staircase and the two men came. The two men and Rob built the roof. They had tied ropes around their waists as the roof was twenty feet high. They did a beautiful job. The lodge looked like a huge home sitting in the middle of paradise, surrounded by Cedar, Hemlock, and Fir trees.

One day while we were walking around the front yard, I said, "Rob, let's have lots and lots of flowers." He said, "I will build a big deck in front and it will hang over the water." I can put flower boxes all around for flowers and the boxes will be wide so people can't fall over them into the water. I was ecstatic thinking about the deck and flowers.

Our next project was our kitchen. Rob would be building a pantry with floor to ceiling shelves. There would be a space for two refrigerators. The pantry was a walk-in pantry. It would be a six by twelve foot room. Rob cut a slab of cedar for the kitchen counter. It was 10 feet by 2 1/2 feet. It was a beautiful piece of wood. The kitchen would have two stoves with a full grill in the middle. I asked Rob to build me a island a couple of feet in front of the stoves and grill. I wanted to be able to turn around and reach my spices. He built this and it was so helpful. He put a shelf across the top, and I put jars of spices on the shelf. I arranged them so that they were all in alphabetical order. This made my cooking so much easier. There were open shelves across the bottom of this Island for pots and pans. On the other side of the island was a bar with a sink. There was a shelf on the bottom of this for liquor bottles. The island was so serviceable. I knew I would be cooking many meals and had to work fast. I did this with this wonderful island. While Rob was building all of this, I was on the beach gathering rocks. I carried up the rocks and piled them in a big pile. I was also collecting them for the rock walls I was going to build, under the windows of the basement, on both sides of the door. The door had little panes on the top to match the side windows. Rob designed this and it really looked great. He would put wood under the windows and I would build rock over the wood. This would be warmer in the winter and cooler in the summer. When my pile was high I went back to work in the house. Rob was cutting the top counter for the kitchen. I started building walls on the floor for the dining room. He came over and we got them done and up. The counter top with the set-in sink was beautiful. He had cut the holes out with a chainsaw. The Health Department said we had to have three sinks and they had to be stainless steel. One sink was for washing dishes, one was for the disinfectant, and the other for rinsing. The whole thing looked beautiful. Everything was coming together inside the lodge. It was time for the sheetrock to go on the walls. It would be a lot of hard work. We were going to get this done by a pro. Rob made arrangements for a couple of men to come and do the work.

Chapter 16

Rob wanted me to go to California and take our furniture out of storage, rent a U-Haul, and drive it back to Campbell River. I flew out of Campbell River in a small plane to Seattle, and on to Sacramento. Arlene and John picked me up at the airport. I stayed at their home for a few days. I wanted to be sure I saw the rest of the family. Arlene had them over one night for dinner. It was so good to see them again. The little ones grow so fast, I had to ask who some of them were. We have a wonderful family. We are all so close—and filled with love. I wish all families could experience this. Barbara and Edward had us over for dinner. I got to be with the family twice. It was great. They wanted to come for a visit to the Island. I told them the furniture would soon be in, and they could come this summer. One of the kids asked, "Auntie, how are you taking the furniture back?"

When I said, "I am renting a U-Haul truck." He said, "Have you ever driven a big truck before?" "No," was my answer, "But I can do it." I always have loved to drive. I have driven since I was fifteen. I have never had an accident or had a ticket. I consider myself a good driver. My dad taught me to drive his pickup. He drove to down town San Francisco, parked, got out, came around to my side, and told me to get behind the wheel. We were with the Cable Cars, hills, people, traffic—I learned how to drive. I was so busy driving I couldn't think of anything else. I thank him all the time for that. I'm not afraid to drive anything or anywhere.

Arlene and I went to rent a truck the day before I was to leave. The men at the U-Haul were great. They found me a truck that was their "best." It was an automatic. It had heat, air, new side mirrors, it had it all... however, it was big! The biggest one they had and I needed it to be big. I said to myself, "Okay Carolyn, you will do this."

When they drove it around from the back, my sister looked at me and said, "Carolyn can you handle that?" The men said, "Sure, she can." I laughed, "No problem." I thought, Jesus, we are going on a trip.

Arlene and I drove the monstrous thing to John and Arlene's house. I stopped the truck in front. Arlene got out on her side, and said to me, "Mary next door said we can park the truck in her driveway, so you back it up I'll direct you." As I looked around, there was John standing on their lawn. He had his arms folded across his chest, legs apart, and a large smile on his face, as if to say, I've got to watch this. I am watching Arlene. She is waving me back, we're looking at each other. She is saying, "Come on, come on," at one point I said, "Are you sure? It looks like were getting close." "No," answered Arlene, "You have lots of room." Then came the crash! I hit the brake and yelled, "Did I hit the door?" Arlene and John are laughing. I am out of the truck and Mary's garage overhang had a rain gutter hanging from the corner. The gutter was ripped off. John says, "A good thing you were going slow, didn't you see the gutter?" I laughed and said, "I couldn't even see the house, the truck is so high." I had to say, "I thought you two were giving me directions. What were you looking at?"

We teased each other all evening. John said, "I'll get it fixed."

We took the truck to the storage place and put everything in it. We couldn't believe it all fit. The truck was full to the door that pulled down in the back. We put the lock on and said, "We are done." I drove back

Secluded Rendezvous

to Arlene's. John had just came home from work. We opened the truck to show him. He could not believe the load. We went to Mary. She was a dear old lady whose husband passed away. Arlene looks after her. Mary laughed and said, "I heard something bang. Take me outside to see. When you live alone it's good to have a little excitement once in a while. I wish I had watched it." Of course I felt bad and didn't need that kind of excitement.

The next morning I was showered by five. I went into the kitchen and John and Arlene were having coffee. John said he had made hotcakes. They are hot in the oven. While we were having breakfast, I asked Arlene about a suitcase with my luggage at the front door. I said, "Did you fill that up with things to take home with me?"

She replied, "You don't think we would let you drive that big truck all by yourself to Canada. I'm going with you, and we'll take turns driving." I was so relieved and happy. I really wasn't looking forward to driving that huge truck alone. I knew it would be a long, slow trip. We had breakfast, gave John a hug and kiss, cried, and were on our way. John was such a dear man. Have you ever met a person who never had a bad thing to say about anyone? That was our John. I met him when I was eleven years old. He and Arlene had a good life together. They had one child, Michele. Missie was married with two children. They were going to look out for John while Arlene was away. I started the drive. I drove all the way to Oregon. Half the way in Oregon I was getting tired. We went to a restaurant. Arlene had filled a cooler with food for our trip. She also brought decaffeinated coffee. We had been drinking coffee along the way. We went to the restaurant to get what she called "real coffee." She told me then, that

Carolyn Begg

if she didn't drink real coffee she would get a terrible headache. My two sisters are wonderful people that would do anything for you. I was sure blessed receiving these two people in my life.

Arlene and I went to a motel in Albany. In the past I would stay there on my driving trips to California and back. Nice people run the inn. We were up at five. There was a McDonalds next to the motel. We ordered an egg Mc muffin, my decaff, and her real coffee, and were on our way again. We stopped for gas and to wash the windows. Arlene was driving this time. She was quiet and finally said, "Carolyn, would you drive through the border?" I did and the border person said, "Pull over there please, and go inside the office." They wanted us to list everything we had inside the truck. I was making a list for some of the things, but it was hard to remember everything. I gave him the keys and asked him to open the back and look inside. He did and told us to be on our way, and have a safe trip.

We drove down the highway to the ferry. Driving the truck on was a little different. We landed in Nanaimo and had a lovely trip along the shoreline. Arlene loved it. I drove so she could look at the water. She loves it as much as I do. The weather was lovely. We stopped along the shore and had a bowl of clam chowder. We both love sea seafood. There is an old fishing boat on the beach used as a restaurant. It is a fun place. We drove into Campbell River. We went to Tim and he tried to set a date to bring our things out to us. He said it might be a month. I told him we were not quite ready yet. We left the date open and I told him we would call in a week. They unloaded the truck. We would take it back in the morning. Arlene and I were happy to let someone else handle the load. We took a cab to the motel and flopped on the beds and took a nap. We were too

Secluded Rendezvous

excited to sleep too long. We took showers and walked up the main street. It was the only shopping street in town. I took Arlene to a restaurant that had home style cooking. We had a very nice dinner. Later we walked to see all the boats tied up at the pier. She enjoyed this. We went back to the motel and called Rob. He was happy to hear from us. It was nice for him to know that everything was done. He would meet us at Heriot Bay the next morning. Arlene and I had a good sleep.

In the morning we went back to the restaurant and had breakfast. We took the truck back to U-Haul. We walked on the Ferry. The water was a little ruff, but not too bad. Arlene held my hand a couple of times. She wanted to stay on the lower deck. We held the rail and watched the water. We took a few sprays in the face, let out a yell and laughed. It was fun. Rob was there with the car when we landed. It was nice. We didn't have to take a cab to Hariet Bay. Rob said, "I'm taking you girls to breakfast."

We went and ordered coffee. We then told Rob we had eaten breakfast. He said, "Why didn't you tell me?" I said we were sorry but we didn't want so spoil his fun, and anyway we needed more coffee. We walked down the dock, loaded the boat, and were home in an hour. Arlene was thrilled with the ride.

When we rounded the turn she saw the Lodge, she couldn't believe her eyes. "I can't believe you two did all this, how did you get way up there?"

I said, "I'll show you tomorrow, when we are on the scaffold."

"What's a scaffold?" she asked. We unloaded the boat and we took everything in. When we walked in she was speechless of it all. "What can I do to help?"

"We will see tomorrow." I replied. We walked all through the lodge. Rob had hired a man to help him with the electricity and plumbing while I was gone. I was so pleased with everything. I knew nothing about these two things. I did have my nose in every wire and pipe. Poor Rob had to explain everything to us. Her mind and mine were very much alike. Everything was interesting to us. The water was to come from a tank we would build later, and the electricity from a generator later too. We stayed there a few hours showing Arlene everything.

We carried the luggage up the bluff and to the house. Arlene loved the woods, the little house, wood stove, pantry, wringer washer, yard and garden, even the outhouse. She didn't expect to see wall-to-wall carpet in the house and outhouse. She loved it all. We all went to bed early. In the morning I was up at five, and there was Arlene all dressed and ready to start helping. I told her, "Coffee and breakfast first then we'll go down to the lodge to work." She was so cute. Rob came in the kitchen. He and Arlene drank their coffee as I made oatmeal and baked bread. She made her own bread at home too. She loved watching it rise in the warming oven, as I did too.

We drank and ate, then started down to the lodge. We guided Arlene down the path. In places it was steep and narrow, and a person could fall off the edge. She was such a trouper. She had no trouble in a few days. When we arrived, Rob said the sheet rock guys would be there in three days. I told him the scaffold is up and Arlene and I will put up the insulation in the walls. I had done some before. We needed to climb up the scaffold. I started to climb. When I was half way up I looked for her and she was looking up from the floor. She had tears in her eyes and said, "Carolyn I can't go up there, I'm afraid."

I said, "Honey, I need someone down there to hand the sheets to me, so this works out just fine." We got them all up and in place. I just had to go part of the

Secluded Rendezvous

way down to reach them. It went fast. She kept saying she felt so bad. I told her if she was not here, I would have to go clear to the floor. She was a great help! I felt sorry for her, I never knew my big sister was afraid of anything, and I told her so along with, "Now I know you're normal." I made her laugh. When I came down, we went to work on the lower walls. It went fast with two of us. We both went up the hill and took an outside shower. Arlene loved it. It was a nice shower in the trees with the birds singing. I told her about the helicopter hanging over me one day. I was so embarrassed I put my bubbles all over me. I bet he had a good laugh. Ollie (my name for Arlene) and I walked to the garden and picked some greens. She was a good cook and said she wanted to cook dinner for us. I got the stove going and started the bread. We had chicken in the freezer and had a delicious dinner cooked by Ollie. I sat by the stove and had tea-how nice! The sheetrock guys called and were coming before noon tomorrow. Rob said, "You and Arlene can stay home tomorrow. These guys will be all over the lodge."

I yelled, "Oh, goodie, we will be there." We all went to bed early. We had breakfast and Rob went down the hill.

Arlene and I were going to see Ken and Marlene. Marlene had always wanted to meet her. "How will we get there," asked Ollie.

I told her we would take a nice walk. We walked down the road. We were quiet. I stopped and said, "Honey, listen."

Arlene whispered. "What's that sound I hear?" I said, "That is the wind going through the leaves, isn't it beautiful?"

Arlene said, "Why haven't I heard that before?"

I said, "Because of the other noises around you." Out here you hear what nature provides." She wanted to know if I came there very often. I said, "Before we were working on the lodge, I came here every time I had a chance to. I came and sat right here. I came when I was alone, or sad, or happy, or missed our family, or when I wanted to pray. Sometimes I just wanted to hear that sound." Ken and Marlene knew I came here. He asked me one day, "What would happen if you died in there?"

I said, "Ken you, and Marlene would know where I was and find me, and you would smile because you would remember me saying, 'Ken could you think of a more beautiful place to die?'"

Arlene said, "I would miss you, don't think that way." I said, "Ollie, think of the things we will get into when we're up there." We continued on our walk. I saw the boat at their dock and knew they were home. Marlene answered my call. We did this on the island to let people know we were coming. "Man or beast?" Marlene was so happy to see us. We all sat by the stove (typical for Islanders) we had tea, they had wine. We all had a great time. We talked for three hours. Ken had told Rob and me about their spot to get clams two years ago. We talked about this and Arlene wanted to go. I told her I would take her tomorrow, and we would have them for dinner. Ken and Marlene were going to town to see her mom and dad, so they could not come to dinner.

Ollie and I walked back up the road and I asked Arlene if she wanted to meet Judy. She said, "Sure." We kept walking and when we got to her gate, Arlene said, "It's so strange to be in all these trees and think your place is the only one here and these houses are here."

I said, "These houses, seem to grow out of the ground."

She said, "Yes, it's so quiet you think your alone. I am glad you are not."

Secluded Rendezvous

Judy had us in, and she was happy to meet Arlene. We had a nice chat, then left to fix dinner. We had a lot of leftover food, so when Rob arrived we were sitting on the deck. Arlene was asking, "Who built the pond, who built the bench?" She still couldn't see how we did these things. "How do you know what to do?"

I said, "I take it one day at a time—what shall we do today? Sometimes it works. Rob and I work together. One of us gets an idea and usually we do it. We look at something, talk about it, and picture what it will look like, and when you have talked it out, you begin to build it. Sometimes you do it together, if it's a big job, or sometimes you work alone. We never said, I don't like that, or no to anything. Rob and I have the same taste, pretty much. I was the designer in the house and yard and Robbie designed the building. We talked everything over before we did it, and worked together." We went in for dinner, and ate by candle light. It was nice. Ollie and Rob had wine. My cardiologist told me liquor was not good for me, so I drank very little.

The next morning we all headed down. Arlene and I were anxious to see what the sheetrock men had done. Rob and I knew we would have some sanding to do, but it would be okay. We would know more after another day of drying. They were now finished with the kitchen. It all looked so good. The place looked so much larger. They were working so I asked Arlene if she wanted to go clam digging. Of course, she said yes. She looked at the boat and said, "Could you drive this?"

"Sure honey, I drive it all the way to Heriot Bay." I took it around the end of the south bay and pulled it up in the sand. We got out with our bucket and shovel. I let Arlene dig. She turned over the sand and had a big smile on her face. She looked at me and said, "Are these clams?" She and I loved seafood.

Carolyn Begg

We filled the bucket and on our way back to the lodge, I pulled the boat over to some rocks on the side of the Island. I drove next to the big black mass hanging there and asked, "Do you know what these are?"

"They can't be muscles, can they?" Ollie asked.

"I told her I didn't know either when I first saw them, we had never seen so many growing together like that." She wanted to know if they were good. I told her she would have them for dinner before she left. We will do this, another day. I tied up the boat. Arlene sat in the boat and looked up at me and said, "Carolyn, you are the little sister. How can you do all these things I can't do? I taught you everything, since you were four." My answer was, "You taught me well, more than you realize. By being the little one, I had the advantage. I watched and learned, from daddy, Barbara, and you. Because I was the youngest, you all felt you had to teach me, and you did. Good job guys! I was taught to love, care, and listen, and learn. I got it all. Thank you. Now give me your hand, and I will help you out of the boat." I walked up the ramp carrying the clams. Arlene was behind me with the shovel, I yelled, "I love you sister, and hurry—we have things to do." She put her shovel down, held on to the ramp and laughed. I was really going to miss her when she left. We went into the lodge, and checked on the men. We asked them to come up with Robbie and have clams with us. Arlene and I got them soaking in oatmeal. We sat on the bench and I told her about the frogs, Freda and Freddie.

Arlene was leaving in two days and we, neither of us, wanted to talk about it, but we did. Rob and I had ordered a plane to come in for her before I left for California. Arlene wasn't sure about flying in a float plane. I told her I loved them. "If the weather gets bad, they just set her down in the water." "She thought

Secluded Rendezvous

that was okay, and liked the fact that the ride was only twelve minutes, to get to Campbell River. She asked, "How does the plane land with those things on the bottom?"

I told her there is water across the back of the airport and "those things" were the pontoons. They allow the plane to float. The plane from Campbell River was flying her to Seattle. Then she would get on a plane to Sacramento. All of this was a bit too much, until I said, "Johnny will be there to meet you."

Then she relaxed and said, "I'll be okay." We went in the house and started dinner. It was fun cooking with her. It was fun doing anything with her. We were very much alike, in many ways, and got along really well. When the men came up for dinner, she asked them how far they got today with the sheetrock. One guy said, "Why, do you two want another day off?"

"No," said Ollie, "We want to work."

The boss replied, "Well, Ollie, you two can return tomorrow, we are finished. We will be sorry to leave this heavenly place, but happy to get back home." The dinner was fun lots of drinking wine, and laughter. There was a lot of admiration about the work Rob and I had done. It's funny, we don't see it as other people do. We say okay, let's cut a tree, make some lumber, and build this or that. Then we go on to something else, day after day. We don't take the time to stand back and say, wow! Look at what we did. We say that's done now we can do this. However it is very nice to have people comment and admire your work! The men thanked us for the dinner, took their flashlights, and left for the lodge, for a nights sleep, and in the morning left for home.

Arlene, Rob, and I went down to the lodge, the next day, to look over the job. The crew had done an okay job. They said it took longer than they thought, therefore wanted more money. Rob gave it to them. We looked over the sheetrock. The seams all had to be sanded so we three got to work. Arlene and I did the bedrooms. Rob did the living room and dining room. The work went fast, but there was a lot to do. We worked all day and finished! It was a boring job, and we guessed the guys we hired didn't like it either. It's too bad, because we can't recommend them to anyone. Their work was inadequate. I asked Arlene if she wanted to take a ride in the dingy, she smiled and asked, "Now?"

I said, "Let's go Rob wants to get wood in I already told him we were going, I knew you would love it."

We got in the little boat and went around to the other end of the Island, where the clams were, when the motor stopped. I could not get it started. I played with the plugs, there was plenty of gas, it just wouldn't start. "What shall we do?" Arlene wanted to know. I gave her an ore and I held one. I taught her how to row. We did this almost all the way. It was hard for her. I said I could do it alone, and of course she said no, so I said, "No problem." I tied all my ropes together, they were long. I rowed over to shore and got out. She yelled, "What are you doing?" I was just making light of the whole thing with her. She was so worried. I told her I was going to pull her to safety. I started walking along the beach pulling the boat. It was easy, and fun. Arlene was laughing. What she didn't know was because we were barefoot, the barnacles were cutting my feet! Rob was outside on the dock watching us. He was about to get the big boat to come look for us. We were all laughing until Rob looked at the blood on my feet. Arlene felt bad, until I handed Rob the rope and went to stand in the saltwater. I told her it was cold and the salt would heal. She thought that was okay, we all had a good laugh. They healed in a few days. The trip was fun and worth it.

Secluded Rendezvous

Ken and Marlene came by to say goodbye to Arlene. We had tea and cookies that Marlene had baked for us. Ollie had swept all the rooms, so we sat around on the clean floor. Arlene announced, "John and I were talking on the phone last night and he said he would come here this summer." We all cheered! John was in Korea during the war. When he came home he said he would never leave home again—not even for a vacation. We were so surprised and happy to hear that he would come.

The plane came in early the next day and the pilot found us waiting on the dock. We were not happy. He took one look and said, "The red eyes lead me through the mist." I introduced him to Arlene. I could tell she was comfortable with him. We hugged and cried, and she got on the plane. We waved to Arlene and the pilot headed towards Campbell River and turned. He flew around the Lodge in a circle. We saw Arlene waving in the window. I think we waved until they landed in Campbell River! Rob put his arms around me as we walked up the ramp. Ken and Marlene had left before the plane came. When we got to the deck, we turned to look out over the water and there was Ken and Marlene waving to us. We knew they wanted to leave us alone with Ollie, but stayed out there to see her leave, nice people! Rob and I got to know the pilots over the years. When one pilot passes over the lodge, he tips his wings to say hello to me. Sometimes when I'm driving the boat out in front of the dock he does this. He makes my day! We were sad with Arlene leaving, however, she and Johnny will be here soon. We'll have to see if Barbara and Edward can come too. It will depend on Eddie's health. Rob and I went inside. He held my hand as we inspected the rooms, and he said, "There ready to be painted." We looked at each other and Rob said, "Let's call it a day and go home." I was glad. I wasn't in the mood to start working again today. Tomorrow will be fine. We walked to our house. Rob started the fire and I made hot chocolate. We talked, listened to music on the radio, and read. It was nice and relaxing. We were in bed by seven.

Chapter 17

The next day it was an early breakfast, down the bluff, and into the lodge—it was so quiet and empty without Ollie. I went into the room where we had our tools and supplies and got out the paint can, and what I needed to paint. I told Rob I would start painting. He said, "That's great because I want to put the porch on the front, then the stairs."

I knew this had to be done now, because the load Ollie and I brought would be here in two weeks. I was thinking "we" would be painting. Not that I minded, but five bedrooms, three bathrooms, a living room, dinning room... I stopped myself and said, "Don't count, just get started." And I did! I painted non-stop in the pantry and kitchen. The areas were small and went fast. I then went to the dining room. I painted the walls as high as I could reach. I took a break and went out to see Rob. Rob was building the front porch. He looked at me and said he was just going to come in for a break to see how I was doing. We went in and had a drink and a bite to eat. Rob had the porch floor half done. The frame underneath was up and looked good. He said he would be done with the floor soon. The railing would be next, but he had to stop and sand the upright pieces. I asked if he would use the electric hand-held sander. He said he would, but there was some area that had to be done by hand. I said, "Have I got a deal for you." We laughed. I told him if he painted the top of the rooms on the scaffold, I would do the sanding. It worked out for both of us. We began after he finished the porch floor. He was pleased with the kitchen and pantry I had painted. While he was doing the porch, I ran in and did the bottom part of the living room. We finished the sanding and painting part we were working on and walked home. It had been a long day. We really got a lot of work done—one usually does when something weights on your mind. We sure miss her! Rob was sipping tea and I was heating soup on the stove.

The phone rang and it was John. He said it sure was good to have Arlene back home. I said, "Who?" John laughed. We all took turns talking on the phone. Rob told John he was making plans for them when they came to Canada this summer. It was good to talk to them and to hear Ollie. The next morning we called Barbara and Edward. We asked them if they could come with Arlene and John. Barbara said they would have to ask Edwards, Doctor, they would love to come too. Edward got on the phone and told me they would be with us. He said if John got on a plane, he had to be there to see it. We were all so excited. When we said goodbye, I said, "Let's go out on the porch Rob. The bread in the oven is not baked yet, and the soup is on the warming side of the stove." We went out and sat on our rocking chairs. The wind sang us a lullaby while the moon's face looked through the trees. We went inside and had our dinner and went to bed.

Early the next morning we were back at the Lodge. Robbie worked on the railing and I painted. I moved into the first bedroom and could reach all the way to the top. It really went fast, using the roller. Rob was doing so well, and the porch was beautiful. I loved it! I always thought a home should have a front porch to sit on and wave to the people going by. Those days have gone by in the city, but here you can, and do, wave to the boaters, and they to you. People love it. Some boaters blast their horn. It's fun! You can also, sit on the porch

Secluded Rendezvous

and watch the whales go by. We saw a baby whale with Ma-Ma and Pa-Pa going by, what a thrill. We worked day after day until the whole lodge was painted and the porch and stairs were done. Tim had delivered the load from California that came in the U-Haul. We needed twin beds so Rob made them. I painted them all white. We were on our way into Heriot Bay, Rob was driving the boat and I was enjoying the water and the Islands. All of a sudden the water was alive with tiny fish jumping everywhere all around us. I yelled for Rob to stop the boat. We saw herring flying through the air and there were seals all around. We looked up to the trees and there were so many eagles Rob and I couldn't believe our eyes. We had never seen so many eagles, fish, or seals all in one place. I started counting the eagles, and after counting four hundred, we drove in to the dock of Heriot Bay. We went into the inn and told them about the eagles. The owner said they come every so many years to mate. The seals drive in the herring and the eagles feed on them. He didn't have an answer when I laughed and said, they were all working together. He laughed and said, " The seals do this when there are schools of herring so they can feed on them. These eagles live around here and come when the seals are here, knowing the herring will be here too."

I said, "I'll be here next year with my camera, I am so disappointed that I didn't have it with me." He told me they have not been there for seven years, they come and go at different times. He said tomorrow there will be so many boats in there you won't be able to go home. We were going to return this evening. We went to Campbell River and bought mattresses. We also used sheets for curtains and spreads to match. Every room was different. They turned out nice. The mattresses would be brought out to us. While in Campbell River we also did grocery shopping, had lunch and went to the ferry. I left Rob and the things we bought on the dock at Heriot Bay. I drove the car to the parking lot up the hill, and walked back to the dock. It was a lovely walk overlooking the bay. I could see Rob loading the boat. We were home in an hour. I started painting the beds and Rob helped. We took the rest of the day off.

It was back to work the next day. The beds were dry so we put them in the rooms. They were twin beds. Rob made slats for the mattresses to set on. We carried everything into the lodge (that had been under a tarp) that Tim delivered. It was like playing house, putting things in place. The dishes went in the kitchen, furniture placed all around. What fun. It looked like a home in one day. We loved it. Rob said "Let's get wall to wall carpet in the living room and hallway."

"Oh good idea," I yelled from the kitchen, "what color?" Rob said, "Let's go next week and look."

When we got into town I would look for dried flowers and ribbons. I wanted to make something for the wall over the beds in each room. I also would mat my photo's, and put one in each room. My niece Leslie had given me mats, and a box of glass to go with them. They were great to have. I was very eager to get going on these rooms. What a fun project. Down the end of the hall was a door that went outside to an area 40x60. The ocean ran along one side of it and the end of this area ran into the sea too. One day I said to Robbie, "Why don't we build a big room here? We could have a bar and a dance floor. I could decorate it with palm trees- Hawaiian style." He didn't like the idea of Hawaiian style. It really wouldn't be right for this area. He was right. However, the idea was there. Rob was very good at planning and he took off on this. It was fun for me to watch his mind work… it was exciting for me. Pretty soon he thinks these things were his idea… and I let him. It really doesn't matter whose idea it was, and he isn't trying to get the credit. He really thinks he thought of it. It's because he gets so into it… it's great. The back of this room could have a big window looking out to the sea. This would be looking toward the Arron Rapids. There is another lodge there called Senora Lodge. Across from it is a lodge called "Big Bay Marina." Nice places.

I went into the bedrooms and began to play with the flowers and pictures. Rob went out the back to rake and dig the 40x60 foot area. He was moving rocks and raking. A big job! I had such a rewarding time. Everything went smoothly for me. We had soup and sandwiches for dinner. The next day we went to town to buy carpet. Marlene and Ken came by and said they were going in to town. They would stay overnight with Marlene's mom and dad. They asked if we wanted to go. We said yes and would meet them on the dock. We were there at the time they suggested and we all went in together. It's always fun being with them. We set a time for meeting the next day. Rob and I got a motel room and went to the carpet place. We found a deep green carpet and ordered it to be sent in the next week. We shopped for groceries, went out to dinner, took a walk on the pier, and went back to the hotel. It's funny, but when you're home, sometimes you say it would be nice to go into town, and when you're in town, you can't wait to get home. I want to be in the quietness of the woods, the neighbor islands filled with the trees, the winds, the birds, the wild flowers and berries, and to know you are suddenly alone. There are no people, cars, buildings, sale, signs in windows, horns blowing, street lights waiting to change so you can cross a street no parking lots, motels, and all the other busy places and things to see. When I step into the boat at the dock and it gently rocks, I say to myself, I am going home now. My heart and head are filled with an indescribable calmness that overtakes me. My whole since of being is transformed. I am happy. I am special to live where I live. But deep in my heart I cannot be complete without my son... nor would I ever want to be. We are home now. Ken and Marlene just let us off at our dock. We exchanged our stories of "town" and laughed a lot. When they pulled away from the dock Marlene and I yelled at the same time, "Come over" and we laughed again. Rob and I went inside and sat on the loveseat and one of us said, "What's next?"

Chapter 18

We decided we needed to start building a water tank. This would be a big undertaking. We first needed to find water to run this place from up on top. The next day we walked down the property my son and his friend Lynn had purchased. It was not being used as yet. As we walked down into the brush, the ground was on a slope and I saw a group of lovely big ferns. I asked, "Are we on our property yet?"

Rob replied, "I believe we are, why?" I said, "If we are, then I believe we just found water." Rob laughed, "I don't see any water."

I said, "We're standing on it, see all these ferns, there has to be water under here."

"No," said, Robbie."

I said, "Okay, let's go home and tomorrow we can start clearing for the water tank." We walked home and had a nice dinner. We went to bed early. The next day we cleared the area for the water tank. We didn't have a lot of clearing to do. We finished early and went fishing. We caught fish and had a fun time. We took the fish up and had it for dinner, it was sure good.

The next day, when we got to the lodge, Ken was there waiting for Rob. He wanted to know if Rob wanted to go back to town with him. Ken needed a few things from the hardware store. Rob needed things for the tank. So they left together. I could still hear the boat when I reached for the shovel and started my walk to the ferns. I pulled and dug, and dug, down deep, and I hit water! I was so excited. I said, " Jesus you're here again—thank you." This is so wonderful; I can hardly wait for Rob to get home. He will be so happy.

I got cleaned up and was sitting on the beach when they drove up. Ken yelled, "Carolyn, have you been sitting here all morning?"

I laughed and yelled, "Sure."

He dropped Rob off and said, "I have to get home, the old lady is waiting for me."

"Wait till I see Marlene, I'm telling her what you said." We could hear him laughing over the noise of the motor.

We were starting the water tank. It would be 16 feet long, 8 feet across, and 4 feet deep. We used cement, and tree posts, under the floor. It had to be very strong for the weight of the water. When we finished the floor and walls, we lined it with a special black plastic made for drinking water. Then we built a roof over it to keep the leaves out. I had taken Rob to see the water I hit. He dug some more. He thought it was okay for now. It's hard to tell how much water you'll get until you give it a try for about a week. Rob ran a water line from the hole to the tank. It had an on and off faucet attached. It was slow at first, but it worked well. Rob wanted to build the deck on the front of the lodge. It would extend over the water. He checked the tide book, and one morning he said, "In three days the tide will be right to get sand from Randy's property." We worked on clearing the area for the deck. It was really going to be nice. Rob has good design ideas. I have often told him he should have been an architect. I believe he is one—without the piece of paper. After three days we took the skiff, shovels, buckets, and a rake, and landed on the beach next to ours. The tide was just right. We pulled the skiff up on the sand, and filled our buckets. The sand on this beach had small rocks mixed in it. It was perfect for mixing with cement. When the buckets were full, we pulled the boat out. It was easy because the tide was coming in and the boat was half in the water. You really have to watch the tide when you do this. When the buckets are full in the boat, the boat won't move without the help of the tide it's too heavy. We unloaded the sand on a tarp. The tarp was above the tide line. We made several trips back and forth until till we had enough sand. That day and the next, we made pillars. We just needed a few cement ones. They would be sitting in the sea. They were very tall. The others were short with a log sitting on top. We had rebar placed in these to hold them together. These pillars were the size of telephone poles. This would be a very strong deck. The tide would come part way under the deck. We called it a day and went home to have a shower.

We had our dinner. We were sitting on the porch when the phone rang. It was Barbara. She told us they were coming with John and Arlene. Rob and I talked half the night, we were so excited we couldn't sleep. The next day the carpet man came with the carpet, and it was down the same day. We were thrilled it was so beautiful.

The cement pillars were on each end of the deck. We used cedar trees sunk in cement. We were working on a surface of rock with crevices large and small. These trees were each cut to size. They all had to be peeled. We could not treat them because the preservatives were poisonous to the fish. We had approximately twelve of them. When they were standing upright in the cement, we started on to trees to be laid across. We cleaned, peeled, and measured them out to be used later. We would have help with them, when our family came the next day. Perfect timing. Rob built a pulley system, so the guys would not have to lift the trees.

The next day was very windy. When the plane landed and Arlene, John, Barbara, and Edward came down the ladder, they had a look on their faces that said, "We're glad to be here." John said, "I'm not getting in that plane again." Arlene told me later, the plane was shaking up and down and John was really afraid. He said it reminded him of flying during the war. We all went in and had tea or some straight shots! We took the rest of the day off… It was fun to all be together again. We had a great dinner, sat around the table, and talked. We wanted to know what everyone was doing. It was so fun. At about 11:00 p.m. everyone was ready for bed. Barbara, Arlene, and I were up early, having coffee. The men came in from outside. They, too, were up early. We told them of our need for help with the dock and everyone was ready. We had a big breakfast. We did the dishes. It was nice to have my sisters there. We were talking about the kids at home. Nice kids. All of them, we felt very lucky. Ken and Marlene heard the plane last night, and came by. Out here, when you hear a plane land, you sometimes don't know if someone is sick or company arriving. They were happy to see our family again. The women sat on the rocks and watched the men move the trees over our pillars. The pulley worked real well. Rob notched them out and bolted them in. Rob said, "If you don't mind tomorrow I would like to finish the deck, but you guys are done. Thank you for all your help." John told Rob, "We're helping. I want this thing to be done before we leave. I want a real plane to have a place to land to take us back to Campbell River." We all laughed.

Ken asked, "What is he talking about?"

Rob replied, "John rode in an Army plane in the wind yesterday, and didn't like it."

John told his story about the service planes. Ken said, "Don't worry John, I'll take you all in with our boat, that way Rob and Carolyn can go too." Ken and

Secluded Rendezvous

Marlene had a larger boat than ours. That was nice of Ken to offer. We all felt better. None of them were thrilled about the plane ride. They were just not used to them.

The next day Rob worked on the deck. Every 16 inches, he had a 2x8 on edge. The deck was to be 40 feet by 40 feet. It would be a great size. There was to be two long and wide steps up toward the basement, then up to a small deck that ran across the cellar door. On the right hand side the front stairs, go up to the front porch. When you go left (on the small deck) you will find a little path leading to another deck. This will be built later. There will be an entrance to the dining room, on the left side of the lodge. Ollie, Barbara, and I took a walk to our little house. Arlene and I showed Barbara around the island. We walked to Judith's, Marks Bay, and down to Peter and Eddies. With no one being home, my sisters were hesitant to walk around their property. I assured them that we all do this on the island. Our doors were left open. We never went in unless we had an emergency. I told them about the gas in my face. My neighbors could come in our house and make a cup of tea. It was a nice way to live. We watched out for each other if someone was gone overnight. It was so good to be walking over the island with my two sisters. We loved each other so. Pokey was with us. I was telling them I wasn't sure about keeping him with the guests at the Lodge, because of allergies, and people being afraid. Arlene suggested bringing him to California. Missie would love to have him. She adores dogs, and so do the children. "I'll ask her, if you want me to." I told her I would talk it over with Rob. Pokey was his dog. We were all tired when we got home, to the Lodge. Arlene said, "You should see where we walked, all over the island."

We sat on the new deck, it was finished. It was beautiful. It changed the look of the place so much...

I loved it. Someone said, "Let's get in the boat and see how it looks from the water." It was marvelous! I thought, I can't wait until I get it landscaped. The flowers will be everywhere. I told them of our plans for Rob to build boxes all around the deck. Barbara said, "You can send us pictures, it will be gorgeous!"

We drove around the Island and yelled at Ken and Marlene from the boat. They came out laughing and waving and yelled, "Come for dinner tomorrow."

Rob yelled back, "Thank you, see you at four?"

"Anytime," was Marlene's answer. Rob drove the boat in, I tied it up, and we all were walking up to the house when Ed said, "You people have a spare room."

"Sure I said, you want to stay here with us?"

Barbara said, "Edward what about the kids? What about your golf?"

We all laughed and Rob said, "Edward you can come back anytime, you're always welcome."

Edward looked toward us, and very seriously said, "It will have to be soon." We didn't speak until we reached the door. I ran in and said, "Drinks and hors d'oeuvres, on the deck in half an hour."

Arlene came in the kitchen and said, "What did he mean?"

I said, "Honey, I think he is sicker than we know."

Arlene and Barbara took everything down to the deck and I got the hors d'oeuvres ready and carried what I could.

Rob said, "I'll get the rest." The men had gotten the drinks and ice down. We couldn't believe the deck. We were all talking at once when I told everyone to be very quiet. They stopped and looked at me and I said, "Listen to the water under the deck." It was splashing on the rocks.

Someone said, "I never heard that before."

Secluded Rendezvous

I thought what a lovely thing to hear for the first time. I told them about the high tide. When the tide comes in we will really hear something. Lovely too, just different.

John said, "It's like having your own water fountain under your deck." How tranquil is that, it is a giant tranquilizer! We were all getting silly and needed to go in. It was shower time. It's a good thing I don't drink, or there would be no dinner. I told everyone to sit in the window seat, enjoy the water, enjoy you're cocktails. I'll get dinner together. We had dinner and played cards after dinner. Guess who won... not the ladies.

Rob was up early and out working on the deck. I was in the kitchen making breakfast. Everyone else was in bed, but awake. I could hear them talking and laughing. I was happy they were having such a good time here. Rob was putting the boards down on the small deck. He will nail them later. He had the two big steps down; they really looked nice. We all had breakfast. Rob announced he was going out to nail the boards. I said, "I will come help you." My sisters offered to do the dishes. How nice to get up from the table and walk out. It didn't take us long to do the small deck and stairs.

When we finished—and I do mean we, Arlene, Barbara, John and Edward came out and were sitting behind us for about an hour. They were teasing us and saying, "That board's too short." and other such nonsense. I told them we would have been done a long time ago if it hadn't been for them making us laugh.

Rob said, "Yes and we would have been half way to Big Bay by now."

Someone said, "Where is that? What do you mean?"

Rob said he wanted to take the guys for a ride to Big Bay Marina. The tide was right for going in the whirlpools. They could stay for two hours. Then they had to come out before the water got rough. I explained this to Barbara and Arlene. Barbara said now she knows why the three of us were not invited. I didn't tell her about the whirlpools, I just told her about the rough water. I told them we were taking the small boat and getting some clams and oysters. They were excited. "Barbara you are going to love this." "I'll get the buckets and shovels." Arlene yelled, as she ran to the basement. I hollered, "And get another life jacket." I went down to the dock and began putting gas in the boat. We three got in, and headed out for the clams. Arlene told Barbara about our motor stopping out there on her last trip. Barbara said she could just see Arlene sitting there like a Queen being pulled in the boat, and me with my feet bleeding. We had a hard time to stop laughing. When we got to the small beach, we got out, and I pulled the boat up. Arlene had a rock ready to put on top of the rope. She loves doing this—anything, as long as she can be in the boat, she is happy. They both were talking about my driving them in the boat. I laughed to myself, and thought the little sister thing again-these two are such fun. They loved digging for clams, seeing who could find the largest one. We filled our buckets and headed out for the oysters.

The oyster bed was a long way for them, and to me too on my first trip here, years ago. When we got there I stopped the boat a distance from the beach. I shut off the motor. What is wrong, they wanted to know. I told them we are there. The tide is out. Barbara wanted to know where the oysters were. I told them to be careful, and to look over the side of the boat. Don't lean too far or hard on the side of the boat or we may tip over; however, the water is only two feet deep. They looked over the side and Barbara said she had never seen so

Secluded Rendezvous

many oysters. She said Edward loves oysters and he has to see this. We got the rake. I pulled one out of the water and gave them the rake. It was fun to sit back and watch them bring the oysters in. They were having so much fun, but I finally said we have to stop. There is a limit as to how many we can take. We poured some sea water in the buckets and took them home. I suggested we take them to dinner to Marlene. When we arrived home the men were there. We showed them our catch. They could not believe how many we had. Edward was so excited. He wanted to see them in the water. We told him we would take him there. John said, "You girls better get ready to go. " He laughed and said, "Girls."

I turned around and said, "I heard that John." We got ready. We put the oysters and clams in the boat and made our way around the other side of the island to Ken and Marlene.

When we arrived at Ken's dock a "Hello" came from the front porch and there was Ken, Marlene, her mom, and dad, all waving at us. It was a nice surprise to see them. They were dear parents. Her dad would joke a lot with a straight face. If you didn't catch the twinkle in his eye, you could take him wrong. We walked through the trees, and up to the house. We introduced the parents to our family. Ted looked at me and said, "What are you doing here?"

I replied, "We are here for dinner too."

Ted came out with, "Well, we will have to see if there is enough."

I said, "We brought a bucket of clams and oysters, and if you're nice, you can have some too."

Marlene said, "Okay you two."

This kind of thing went on all night with Ted. He was a fun man. We had a lovely time at the table. The men told their stories of Big Bay. Barbara looked at me and said, "You didn't tell me about the whirlpools, Carolyn." I winked at her and smiled. Ken had a marvelous bar-be-que of salmon steaks. Marlene had a delicious pot of beans, a salad, greens from her garden, clams, and oysters. Marlene asked me to cook the oysters. It is a recipe I made up. Everyone seems to like them. Everyone loved Marlene's beer. It was a big hit with the men at parties. Marlene showed Arlene and Barbara the boxes Ken made on her porch. They went around the porch. They were long and wide. Really nice. She planted her vegetables in them all summer. It worked well in the direct sunlight. Marlene had a hot house for flowers and spices. She had flowers across the front of the house and lovely hydrangeas in the shady areas. We enjoyed the walk around the yard. It was starting to get dark, and time to get in the boat. We said our goodbyes and thanked them. We always had a nice time. We were all in bed early.

The next morning at breakfast John announced, "Well, folks, tomorrow is our last day. We have had a wonderful time, thank you. Don't think you are getting rid of us, because we will be back."

At the same time Rob and I said, "No." We were laughing when I added, "I think you should go home, pack, and return for good."

We all had a serious talk about this kind of life. John said, "Carolyn I really admire you and Rob for having the courage to leave your country, jobs, families, and friends."

Edward said, "You two are amazing! The courage you had to do this, the danger, living in a tent in January with snow, and bears and cougars on the other Islands that can swim over."

I said, "Rob, go get the suitcases, we will go home with them." We had a good laugh. My heart was sad.

Secluded Rendezvous

We would surely miss these loving people. How blessed we are to have them. The sun was out, it was lovely. We sat on the front deck. We thanked the guys again for the support and help they gave. No one wanted to go any place this last day. We just wanted to be together and talk. We walked down to our beach and sat on the logs. We told them of our plans of the lodge, how it would work, of the building of two more decks, and the addition on the back of the house. We took a walk around—they wanted to see it all. We went in and turned on some music and sat in the window seat. It was a very relaxing place. The window was facing the new deck and out across the water, then to the island across from us named Raza. I told them the Killer Whales swim by, but not now. They have their own time. It was almost dinnertime, so we had before dinner drinks. Arlene, John, and I used to drink scotch and water. I don't anymore, but we had bought some for them. Edward had to give it up too. He and I were happy with our ice tea. We teased everyone by telling them they would have a big head in the morning, but not us. My sisters and I fixed a salad, rice, and vegetables in the kitchen. The men were outside bar-be-queuing on the new deck. Barbara said Edward was going in for some test when they returned home. We all knew he wasn't himself, but he really tried hard to hide how sick he felt. We put the food on the table and the men came in with the fish. We all loved fish and could eat it every day. I believe we did. It was a nice dinner but quiet. Not like the other nights when we all talked and laughed. Rob and I were asked, several times, if we needed anything, did we have enough to get started, and finally do you two need any money? We answered, "No, thank you. It is a little tight, but we will be fine." We visited for a couple of hours more. Arlene and Barbara had to pack so we went to our rooms. No one wanted a full breakfast. They would have a snack on the plane. We had juice, coffee, fruit, and toast. Ken and Marlene were there on the dock when we came out. We loaded the boat, said bye to Pokey, took him in the house, and left. Pokey doesn't care for the boat. He would rather stay home. We arrived in Campbell River, took them to the airport and came back to Heriot Bay. Marlene talked about them. How nice they were and the good times we all had.

We headed out, in Ken's boat. We were all quiet. Ken opened three beers and passed them around. I had my seven-up. We drove passed a spot on an island. I said, "Ken slow down please."

He did. I said, "Marlene, remember when Rob and I went to town with you, the first time?"

"I guess so." she said. I reminded them of the time it was really blowing and Marlene was afraid and told Ken to pull over and let her out. "I looked at you and asked where you were going. You said you were going to walk. You told me this was still Quadra Island and I can walk to Heriot Bay. Ken will get there before I do. He will get the truck and meet me. I said, 'it is all trees', and you said there was a road back there behind them. Ken pulled over, and Marlene asked if I wanted to come. I told her I wouldn't let her go alone, and we both got out. Now Marlene and I are in the middle of nowhere, nothing but trees. No boats, houses, docks, nothing but trees. Ken angrily backs up the boat and yelled, "We will be at the Heriot Bay dock meet us there."

I yelled, "What time?" Marlene said to me, "It doesn't matter, he will meet us on the road with the truck. He always does." I asked you if you did this often and you said only when I don't like the water. Marlene and I took a very long walk through all these trees, but it was lovely, after I got past the worry of bears

Secluded Rendezvous

and cougars. Marlene said, "Very soon, you'll see a lovely house and yard." We walked up to this ten-foot high, wire fence. The yard was at least one acre. It had a few trees in it and the space was covered in a rich, thick, moss. There was not a weed to be seen. Marlene said, "This man was in the newspapers. He planted this. It took years to get it like this. He had trouble with the animals, so he reluctantly put the fence up." He had Azaleas, Hydrangeas, and Rhododendron, along the inside of the fence. One can't imagine how peaceful it was. Ken and Rob came alone in the truck and picked us up. I had told them I was worried about the bears and cougars. Ken said I should have been. Just because we didn't see any, doesn't mean they are not there. He told me he has seen tracks on Rendezvous. He said there haven't been many. The way we all walk around the island, we should all carry a gun. I told him I had a hunting license. He told me I couldn't kill anything with a license. I told him I thought if I just showed it to them, they would go away. I got a hard look from him with a smile. I said, "Just keep your eyes on the water, and we will be home soon."

Ken and Rob jumped out of the boat we had stopped at the post office. I told Marlene her husband was a nice man, and fun to tease too. I told her I hoped she didn't mind. She assured me it was fine, she was used to it because of the way her dad teased. I told her my dad taught me. The men got back in the boat and Rob said, "Open this from my mom, I want to see what this is." Rob was reading a letter from a man that writes for a newspaper in Vancouver. He wants us to call him. He and his son want to come to the lodge. Everyone was excited. They would be our first customers. We reached the dock and they let us out and wanted to go home. It was a long day for all of us. We thanked them for taking us, and our family to the airport. They said that is what friends are for. As they drove off I said how nice they are. Rob said he offered to pay for the gas and Ken said, "No, I'll get it back one day," and laughed. So that's what we did over the years with most of the people on the island. We helped each other.

We went in and laid in the window-seat and as I fell asleep I thought of my sisters. We slept for two hours. We woke up and Rob said, "Let's build decks." We went out on the side of the house and I began to pull and clip the Lemon Leaf, they had tough roots. Rob was measuring and cutting boards. When I had everything cut and pulled, I went to Rob to see how he was doing. He said it was going slow. I told him to measure and I would cut the boards. We had done this at the little house and it worked well. I liked running the saw. When we had the boards cut he needed, we went to where I was working earlier. He said, "I'll rake and you put the stuff in the wheelbarrow, we'll take it around the back where it's clear and burn it later." We mixed cement for pads. When the pads were in, it was getting dark, we called it a day. We were tired, but thought we might start having customers soon. We had to get this deck done. We had canned soup for dinner. We occasionally do this. We don't mind. We enjoy it sometimes, and it sure beats cooking every night. Rob said we would do the little kitchen deck later. No one goes out that way but us to check the water tank. We were in bed right after dinner. We talked about the deck and how we would work in the morning.

Chapter 19

We had Oatmeal and tea and went outside. Rob was going to fit the boards on and I would help him nail, and we did this, and we finished at dinnertime. The next day we were sitting in the window when the phone rang. It was Arlene. Mary, her neighbor, had a couple of friends that wanted to book a weekend with us. There was a phone number for us to call. We called and they wanted to come for a weekend and two days. We booked them arriving Saturday and leaving on Tuesday. We hung up and said they will be here next weekend. We went all through the house and made sure everything was in order. That night at the table Rob said, "Shall we dare put the flower boxes in, can we do it in a week?" I said, "Robbie, we can do anything—but should we? Let's go for it. I'll bet there is enough wood. It just has to be measured and cut." Rob got up and started for the door the next morning. I told him we had to have breakfast first. I knew we would not stop for lunch. When we finished breakfast we went outside. Rob did a walk around and said, "Let's plan this" and we did. There would be two boxes across the front. One on each side of where the ramp would be. There would also be one going down the two sides. These boxes were to be knee high and wide. It was for the protection of the children, having them wide, they couldn't try to look over the side and fall into the water. Rob started building. I got my clippers and rake, and shovel. I started on one side and went all the way around the house. I have been wanting to do this for a long time. Everything was there. The Lemon Leaf, ferns, rocks, tree stumps. I wanted to plant in the stumps. They were lovely with moss growing all around them, all over the bark. What a delightful time I was having. I landscaped the whole yard. I went back to Rob and he had finished nailing together two boxes. They were wonderful. He said, "Help me put these in place." We put them across the front of the deck. What a difference it made. You looked over them to the water. Before you looked at the sharp edge and had the feeling you could fall off. Rob was starting to work on the smaller box for the side to the tree. I asked when we should think about getting a ramp. He said he would call Mernie tonight. She had told him she knew someone who made them. He builds them out of steel and lives in Indian Arm, so it will have to be shipped. It sure would be nice to have one for the customers. If not next week, the pilot can land next to the float. I can bring them in to the beach in the small boat. Rob finished the box and came in for dinner. I had it ready and we ate early. We were hungry. After dinner he said he would build the last box tomorrow. He made a call to Mernie. We both talked to her. It was good to hear her voice. She also lives in Indian Arm. She said she would call the man and call us back. She did and then he called us. He was working on one and would call when it was finished. We wouldn't have it for our first customers, but soon. Rob had all the boxes finished by noon.

The next day he trimmed all the boxes. The end of the long box was six inches higher than the rest. It really looked good. I told him we now needed dirt. He laughed and said it was in the basement. I turned and opened the door. He said he had been wanting to get rid of some dirt. He said he wanted to dig the dirt on the sides lower. This was so he could stack wood on the sides. We got the shovel and wheelbarrow. We had lots of dirt, but it had rocks in it. He took a big piece of screen and put a frame around the edges. This was laid on the wheelbarrow and dirt shoveled on top. It worked well. He brought me the dirt and I sifted it and put it in the boxes. The rocks were sitting on top of the screen. I took the screen off and shook the rocks in the water. It worked really well. We worked on the boxes the rest of the day. When we finished, we went in and had dinner. Rob looked at me and asked if we should take the boat in and get some plants tomorrow. Of course I got very excited. There are two nurseries in town. We can go early. We did go early and had a fun day. Most of the town trips are hurry in to get something needed now. Then hurry home to use it. The trip in town was relaxing. We went to the store for fruit and fresh greens. We put these in the car and admired all the plants in the trunk. They were beautiful, all kinds and colors. I suggested we have lunch at Heriot bay. We ate outdoors on the patio, it was very nice, sitting there talking to the folks around us, and looking out over the water. We finished lunch, took the car down to the dock, unloaded the car and filled up the boat. I drove the car up the hill to park. We were home in forty minutes. Rob got the plants out of the boat. I started carrying them up the ramp. It looked like we had as many plants as the nursery. I started planting. What an experience. I have never planted in such an open, spacious, peaceful setting. Every plant was set in the box for it's height, size, color, or scent. When our furniture was put on the dock we put a small table with two chairs aside for the deck. Rob brought them over. We put them in a corner where two people could sit alone near flowers and overlook the water. It was perfect. I placed the flowers with a lovely scent there in the box. We bought two climbing roses. Rob was building a trellis to go over the top of the ramp. One rose would go in each box on each side of the trellis. We had eighty plants. The planter boxes would be very full and any flowers left over would go in planters close to the house. Wherever we could put dirt, there would be a plant there. The flowers were all planted and it was a spectacular sight. Flowers give so much beauty to any area.

Chapter 20

Our first customers came and stayed three days and went. They seemed to have had a nice time. We enjoyed their stay. The writer from Vancouver and his son came the next week. He was a very nice and interesting man. His son was a delightful boy. We all had a fun time together. He asked us all about our experiences on the Island, and was taking notes. He and his son did a lot of boating and fishing. They loved the lodge and of course, the area. About a month after they left, we received an envelope from him, in it was a newspaper with a story about us and our Lodge. How very nice of him. We had a few customers after, but not really enough to get this business going. We needed to advertise more widely. I suggested we get a brochure made. Rob said he would make one. He made a beautiful brochure.

One morning Rob suggested I take the brochure to California and have them made there. Pokey and I got in my Torino and headed for California. I love driving this car. We took good care of each other. I was on the highway one early morning. An old truck with medal sides was in front of me. One side flew off, and landed in front of me. I took my foot off the gas, checked the traffic around me, swung around it and was in the lane of the on coming traffic. I rode the car around in a circle till it slowed enough for me to continue in my lane. I knew Jesus was with me. I am not afraid to drive this car anywhere. It had a 429 engine and if I have to have the power, it doesn't hesitate a second, it just goes. I do use it sometimes. Pokey and I stopped in my usual place in

Oregon. Pokey was so good. No barking. We were on the road and arrived in San Francisco in the afternoon. I stayed with Randy. I told him of my plans. In the morning I took the brochure in to have them made. They did a fantastic job on the color. I had boxes of these made. I put them in the trunk. Good weight for my car. That night my son, asked me where I planned to go. I said Los Angeles and he said, "Why, don't we go to Santa Barbara too?"

I said, "We?"

"I'm coming with you mom, I took a couple of days off work. We can leave in the morning for Los Angeles. Kathy and the kids want to play with Pokey." I was so happy. How very special my son is! I fought off the tears and said, "Everyone should have a son like you, I love you Randy. Thank you." "I love you too mom!"

We left in the morning and arrived in the afternoon in Los Angeles. We stayed overnight and made a list of Travel Agencies. Randy drove and he and I ran in. We had no time for appointments. The people were all very nice. I gave them a short talk about Rendezvous and left a stack of brochures. We lost count of how many places we went to. We did this for two days, stopping only to eat and sleep. We had so much fun too. It was nice to have him do the driving. I got to see the sights. Just to be alone with my son was a blessing. It had been such a long time. We drove to Santa Barbara. We distributed more there, and started driving home to San Jose. Randy drove the whole trip. He said, "You still have a long trip back to Canada, mom." We arrived home late, had tea, got up early, Randy went to work, and Pokey and I left for Sacramento. I was very worried for Randy, he had told me he and Kathy were getting a divorce. I was sad for them all. In an hour and a half, I was at my sister Arlene's door. She was waiting for me. She and

Secluded Rendezvous

I were going to the Travel Agencies in town. We first stopped in where John worked. It was so good to be with them again. John gave me a phone number of a man named Bell. Bell had a radio show for fisherman. He would take phone calls for information all about fishing. He would tell them where to fish, what kind of bait they should use, and answer questions about fish. John said to use his phone and call Bell. I did, and Bell gave me a time for Arlene and I to go in and see him. We did and he wanted a package of brochures. He was very excited about the lodge. He said he and his wife would come to stay with us, and they wanted to bring a couple with them. He would call his friend and then call me at home. Arlene and I were so excited. We handed out many brochures, were in and out of places, and got tired. We got a great idea. We went to a stationary store and bought a box of large envelopes. We got the phone book and started mailing brochures. We also did San Francisco. Missie, and her two children, and Barbara and Edward came over for dinner. Arlene was a good cook, so we were all happy. I'm always happy when we're together again.

We called Rob, and told him what we had been doing, and that we missed him. I told him I would be home in a couple of days. I would be leaving in the morning. Randy called to say he loved me. We never say goodbye. We say I love you! We had a wonderful dinner, and a nice evening. Rob said it was okay for Missie to have Pokey. She loves dogs and we knew he would be happy with Missie and the children. He was getting old and pretty soon it would be too cold for him in Canada. I got up in the morning with Arlene and John. We kissed, hugged, and cried, like always, and I left for Canada.

I drove for nine hours to Oregon. Arlene gave me a bag for my cooler. In it were cookies, a baked potato, chicken, an apple, banana and grapes. I had enough food for three days. I didn't go out for dinner. I took a shower and got in bed early and had a good sleep. I was on the road bright and early. I sure missed Randy, and Pokey, too. Sweet dog. I have only six hours to drive to the border. It will take an hour and a half to the Ferry. Mom and Haakon were away when I called, so I kept driving. Sorry I missed them. I got on the ferry and went to Nanaimo, from there I had a two hour trip down the coast. My favorite drive was along the Coast. Campbell River was my next stop. I had called Rob. He was meeting me in Heriot Bay. I was dead on my feet, but okay. I loved every minute of the trip. I love to drive a boat, car, truck, and would fly a plane if I were fit to do so. It was good to see Rob and home. We talked for hours over dinner. I told him about Randy and Kathy, he was sorry, but said Randy would be okay. I told him where we drove, where we stayed, and everything I had done since I left. He brought me up to date on what he did. He finished a deck on the kitchen side. It was a cute little deck. He made it so the path I had made with rock led right to it. The path went up to the water tank. We had dinner, and talked about Pokey. I told him he would have felt better if he had seen Pokey with Missie and the kids. Pokey was used to Rob's girls, Karen and Claire, he liked children. I understood how he felt. He had him all those years.

    The next day a huge ship pulled up and the Captain got on his horn and yelled, "Is this the Begg place?" Rob yelled back and the Captain dropped his anchor. He had our ramp and we were so thrilled.

    The crew asked, "Where do we put this?"

    Rob said, "Our dock is not built yet. I am going to bring our small float to you." He got in our boat and tied on to the ship. The crew put a lift on the ramp and

Secluded Rendezvous

Rob guided it with a pole hook on to our float. The float was taken to the spot where the end of the ramp was to go. The ship pulled out nice and slow, not to make waves. They all yelled, "goodbye, and good luck." He was a real nice Captain, and had a nice crew. The tide was high and we were glad. The top of the ramp would go to our front deck, in-between the flower boxes. Rob attached a heavy rope to the top railings of the ramp. As the tide came in, he pulled the ramp up until it was in place. This took a long time for the tide to be right. We then took a small log to put under the top of the walkway on the ramp. Rob put a jack under the log. As he jacked it up, I rolled the log under. As soon as it was in place, he put pins in the holes he had drilled earlier. The whole front was pinned and bolted together. The bottom of the ramp was being bolted to the small float. When we were done, hours later, we looked at each other and said now, we need a long dock for boats to tie up to and to put the bottom of the ramp on. But for now we had a float with a ramp! No more crawling up the rocks from the beach .

A week later Bell, from Sacramento, called. Bell said his group of four were ready to come. We set the date and they would be here in four days. We called the floatplane Seaport and told them when he would be there. The next day Rob got the boats ready for them. These were real fishermen, and would being using the boats most of the days. I went around the house and checked everything, and everything looked good. I kept it clean and ready for guests at any time. I went into the pantry and it was ready also. Rob chopped more wood and I helped carry it in. On the fourth day we heard the plane coming in and went down to the float to meet them. They were very friendly and nice to greet. They were ready for a "rest and vacation." They seemed happy and relaxing to be with. These were going to be our kind of people. We took them and the luggage into the Lodge. They said, "Look, at this place—it's beautiful." We thanked them and took them inside, to their rooms. It was late in the afternoon when we served drinks and hors d'oeuvres. These we two couples had been friends for years. They told us stories of when they met and of their lives and jobs. Bell had a particularly interesting job with his sports radio show. We went in to dinner after a few hours and talked through coffee. They wanted to hear all about us, and how we arrived on this Island. I suggested we leave our story until tomorrow. They admitted to being tired, Bell had done an early show that morning. They retired to their rooms. I did the cleaning up, and got some things ready for breakfast. I went up to bed. We were up early. Rob started a fire in the livingroom fireplace. I went in the kitchen. We had a long narrow counter in the dinning room. I put cookies, potato chips, bread, lunchmeat, pickles, fruit, and whatever else I thought they might like for lunch on that counter. We had a little cooler with ice for each guest. They got to make their own lunch, and they loved it. It was easy for me too. I gave them a healthy breakfast. They could have eggs, pancakes, or oatmeal, and of course, everything that went with it. There was always homemade bread.

They were taking the boat out to fish and explore the area. They asked what time dinner was, I told them, and they said they would be back at that time. I thought that was great. I used that method with all our customers from then on. Fishermen usually don't like to be told when to stop fishing. I having a love for fishing, understood. Being the cook I also understood my side of trying to keep everything warm and fresh. I never had a fisherman complain they were always home on

Secluded Rendezvous

time. Bell had asked if he could bring fish home for dinner. I told him I would love it. I said I could eat fish every night and Rob too. We had a lovely day! We were through with our work by noon, sat on the deck most of the afternoon drinking tea. This was different for us, as we always found a new thing that needed to be done. We couldn't do this when guests were there. We were enjoying our rest. We did talk about what else we had left to do. We wanted to have everything done before winter. I went in to make bread and salads, and cleaned the vegetables. I went back out to the deck and Rob was measuring a path that led from the front deck to the one side deck. This "path" was to be a ramp for wheelchairs. You would go up and across the side deck, and up another small ramp, and on to a small porch that came out of the living room. This was another of his great ideas. I cried, "Let's put railings in the showers too." And so we did! But, not this day.

The folks came in for dinner, holding a big Ling Cod. This was my favorite fish. They were so excited. The fish was a beauty! They said the fishing was great. They released a lot of them. They were all talking at once. They went into their rooms to get ready for dinner. They joined us in the living room, where we gave them a glass of wine. They were hungry. Everything was ready but the fish, and it was in the oven, and would be ready soon. We went to the table and had our salad. Being out on the water makes you very hungry. We answered a lot of questions, naturally. How does one leave California and find an island when you were raised in California? How did you find this island, miles from town? One of the ladies wanted to know how we got a decorator to come way out here. Rob said, "Carolyn did it all, and she does the gardens too." They thought we had hired people to do the building too. We talked into the night.

The next morning, at breakfast, we started again. What kind of chainsaw did I have? How did you carry all the rocks for those planters? What kind of trees did you use for this and that? Most of the questions were directed to me, and I knew the answers. We knew he didn't believe we did all this together without help. We all left the table—they were getting ready to go fishing when I said to this guy, "I'm going to go out and chop some wood." He turned to Bell and said, "I'll meet you in the boat." He followed me. I picked up the axe. He yelled, "Here, let me do that." He took the axe and started to swing. I yelled, "No!" I took the axe from him and said, "Hold it like this, and bend your knees, bring it down straight or you'll cut your legs off." I brought the axe down and the log split right down the middle. He said, "Wow, okay now I believe you." He turned and went down to the boat. We could hear them laughing as we stood on the deck waving to them. The man looked up and said, "The fish are waiting!"

I answered, "Have a wonderful day, and if you need help catching one, I'll show you how." They all had a good laugh, as Robbie smiled and shook his head.

Rob stayed outside looking at his idea for a ramp. I cleaned the kitchen and their rooms. Rob came in and I asked him to vacuum. I went back in the kitchen and got ready for the dinner. We were having fish again so it was easy. Rob and I later went to the deck with our cups and talked. We were talking about the niceness of this -meeting different people from all over. They are on vacation. They are fun and good to be with. Bell and his people left in the next two days. He said, they loved the place and admired what we had done and us. He said he would be sending us customers. He kept his word. We had customers throuout the summer. It was not as busy as we would have liked, but it was okay.

Chapter 21

The couple that owned Big Bay Marina came to us at the beginning of winter, and asked us if we would stay at their place and take care of it during the winter. They offered us a good salary. We would be closed at rendezvous, so we said yes. We winched our ramp up on the deck, and pulled the float high on the beach. The ramp was hanging over the side only two feet. We left and went to Big Bay. It was only five miles from us, so we could go home once a week and check on our place. We had a little two bedroom house to stay in at Big Bay. The place had cabins surrounded by lawn. There was a bar and a grocery store, a laundry mat, their home, and gas pumps. The gas pumps were open to the public. There was candy and cigarettes to be sold. There wasn't too much to do. Because of the weather, there were not a lot of people buying gas. Rob had to mow the grass, and clean up the branches and leaves from the storms. The wind was wild. It was raining and very cold. The front of Rendezvous took the bulk of the storms. We were in the middle of the island and very protected. This place was opened to the Alaskan winds.

One day a man from the lodge across from Big Bay came to us and asked Rob if he would work for them, as a carpenter. He said yes. I became the "gas attendant." Rob stayed there most of the time, because of the weather. They also had a bar, pool tables, and a hot tub. This began my feelings of being left alone. I spent my days checking the buildings, pruning plants, and bushes, shoveling snow, and tending to the boats. I was really my own boss. I could do whatever I needed to do.

Carolyn Begg

Pumping the gas was easy. There was a sign on the dock and a bell. The sign said, "Ring the bell for service, and I would walk down the dock." The people were usually very nice. Every time I was asked, "Are you here alone?" I laughed and said, "No they are all in by the fire, it is my turn today." I was careful of the animals. I carried a walking stick. They say cougars don't attack people, but they have. Usually you can wave your stick and yell while working your way to the house. They also say to hold your coat up in the air and yell. The coat makes you look like a big person, bigger than the cougar and they are afraid. Jesus was always with me, since I was a child. I would say, "Please come with me" and I would not be afraid. For me, it works. The worst of all, this job was shoveling the snow. I would shovel down the steps from the house, down the walk, across the bridge, down a little slope and on to the dock. The dock had to be done well, for the customers. I did this every day. One time it snowed for two weeks straight. One day Rob came home to stay overnight. This would happen off and on. I moved into the other bedroom. I was not happy with being left alone so long.

He came one day and said he wanted to go to the lodge to get a few things. Would I like to go? We went in the big bay boat. This was a big boat he was driving. We pulled up to the front of Rendezvous. There were some big rocks, even with the boat. Rob pulled up to them. He took the rope and went to jump out to the rock and he slipped and fell down to a small place on the beach. It was rock mixed with sand. I jumped off the side and grabbed the rope, tied it to the end of the ramp, and ran to Robbie. He was in pain saying, "My leg, my leg."

I was kneeling next to him and I said, "Robbie, I have to get you out of here." He was turning very white

Secluded Rendezvous

and his eyes were rolling back. I thought he was going into shock. I opened his shirt and belt, and threw a hand full of water in his face and said, "Stop it and listen to me—the tide is coming in I have to get you up this rock. No one ever died from a broken leg, but you will die if you don't get out of here."

He said, "Tell me what you want me to do." I rolled him over, had him grab on to my belt. I pulled him back to the rock, put his arms straight up and told him to hold on to the rocks. I put my shoulder under his bottom and prayed, and pushed, and up he went. I pulled him up from the top and over. What a relief. I hugged him and told him, "I yelled at you to keep you from going into shock." I asked him to lie real still and began to check his body. I believed it was his hip, but didn't know if it was broken or not. I didn't want to move him too much. The floatplanes don't fly after a certain time and it was too late now. I made us a bed (it was very cold) until the tide was high. I took the blankets and pillows to the boat, from the house. I untied the boat and pulled it over to Rob. I helped him into the boat. He wanted to drive. He said he would feel better sitting and hanging on to the wheel.

We docked the boat at Big Bay. A kid came over to us and introduced himself. He said his name was Wayne. Wayne said he was a mechanic working for Big Bay on the boats and motors. He had been at home with his family and came to do a job. He lived on his boat, and had an ultralite plane to travel home in. While he was talking, he got Rob into the bed I made on the side seat of the boat. I had called the plane and they would be here as soon as it was light. It was better for Rob to be in the boat. When they got here it would be easy to transfer him from the boat to the plane. I was afraid to go to sleep. I had taken a semester of Emergency

Care. I knew how broken bones could cut you inside. I gave him a cup of broth and hot tea. He fell asleep. I had given him a phone. I had one and so did Wayne. Wayne's boat was tied up next to Robbie. We all left our phone on all night. During the night you could hear, "Rob, you okay? Yes, good!"

The plane came at daylight. They flew us to the Hospital in Campbell River. His hip was shattered. The doctor said he had never seen anything like that. He was in the hospital for a week. They had a house next door for families of the patients. I stayed there. Wayne pumped the gas for me at Big Bay.

Ken and Marlene came and picked us up. The time was within two weeks of our being finished with our job. Wayne took over and we were grateful. We went back to the lodge. Rob healed remarkably well. He used a cane. He got around very well, even in the boat. He went back for a checkup and the doctor couldn't believe how well he was doing. We stayed overnight in the motel. In the morning there was a storm. I said, "We will stay another night." He didn't want too.

We went to Heriot Bay and the sea was wild! I told him I was not getting in the boat. Our boat was a seventeen-foot boat. It did not have a cover on the top. Rob said, "Stay here, I will be back." I waited for him at the dock. When he approached me he said, "You'll go home in a covered boat, it is carrying four other people and you will make it home just fine." He told me he was going home in our boat. I asked him to leave our boat there and we could come and get it in our skiff, and tow the skiff back. "No," Rob said. He got in his boat and left. I left with the other boat and it was a rough ride. The water was over the top most of the way. Everyone was drinking beer, and laughing, and yelling out of fear I was sure. I said, "Lord please get us home safe, and

Secluded Rendezvous

Rob, too." I was let off on the other side of the island. It was too windy on our side. I walked over in our direction and got lost. Now, you can't get very lost on a small island, so I finally found my way. I was soaked all the way through.

Rob was home when I got there. The next day the radio said we were in eleven-foot waves. Thank God, summer is coming. I called Barbara the next day, to see how everyone was. Poor dear Edward now has Leukemia. He is very ill, and so brave. If you didn't know of his heart problem and blood problem, you would not know he was so ill. He is going through so much. He's so sweet and everyone loves him. Barbara said he goes to the club, dressed and looking like nothing is wrong with him. She put him on the phone to say hello. We talked a bit. He laughed. I said, "I won't keep you Edward. I'll see you soon, I'll come for a visit." He said, "Dear, Carolyn, you won't be seeing me again."

I said, "Oh, Edward."

He answered, "It's okay, I know it will be soon." I never saw Edward again, however, I do talk to him. I loved Edward he was good to me, and everyone.

Chapter 22

We stayed in the house a lot this winter. The winds were 90 miles an hour one day. Ken drove his boat by and yelled to us, "Get in the house-a butte wind is coming." We were on the rocks looking at the water. All of a sudden we heard a roar, we looked down the water toward big bay and the trees were leaning toward us. The spray off the water was unbelievable. We watched it come up to our place and pass us. The wind was so strong we held each other. It was a great feeling to be a part of it. Some people down by big bay lost most of their trees. The people said the trees went down like matchsticks on the ground. One person lost his roof. One person lost his dock. That winter we lost our water pipes. They froze and broke in pieces. It was so cold we had icicles on the inside of our windows. This was downstairs. We didn't leave the fireplace lit at night. We lived upstairs and had an airtight stove. We banked it and it went all night. We were never cold upstairs. We took hot bubble baths and sat in front of the stove, drank tea, and read our books.

When the winter storms were gone we went to Campbell River to get fresh greens and fruit. While shopping I was talking to a lady who asked where I lived. I told her, and she asked if we had a dog. I told her we used to but we gave it to my niece. Then I told her about Pokey and the customers. She said she knew an old lady who had been left with a puppy and that puppy was in a cardboard box for two weeks. Rob came to where I was and heard us. I told him about the puppy. I wanted to see it. We went by the house and looked at this darling baby cocker spaniel crying in an upside down box. There were no holes in the box. We picked it up and both wanted her. The lady yelled, "Put her back in the box she will wet on the floor."

Rob said, "I'll buy it for you." I fell in love with this puppy. We got in the car and Rob asked me her name and I told him it is Megan. We took her home and I started training her. She was so good! She was beautiful. Rob told me when we got up in the morning he was going to work for a man that ran a fish farm. It was on another island. It would be too far to come home everyday and his schedule was 10 days on, and 10 days off. I was crushed. I didn't understand why he was leaving me again. He said we needed to make money. I asked why we didn't we live in Campbell River. In the winter we both could work and run this place in the summer. He didn't want to do this. I said our original plan was to live in California in the winter and work, and live here in the summer. Why can't we do it? He was going to work on the fish farm and he told the man he would. He said goodbye and left. He would call to see if I was okay. "I guess if I am not, you will find me dead in ten days," was my answer.

Ken and Marlene bought a fishing boat and a license to sell fish. They are fishing every day in the area. When they heard Rob was working on another Island, they said, "Carolyn, we will come by every week to make sure you are okay." I thought that was so nice. Rob was home in ten days and gone again. The time went so fast. I was so thankful for Megan. We went fishing almost every day. She began to love it as much as I did. I never believed in my dogs, in the past, sleeping on my bed. Megan slept next to me (on her blanket) every night. It was a much needed comfort to have her. We were together every hour. She followed me like a

Secluded Rendezvous

little shadow. I loved it. I woke up one morning, walked Megan outside, came back upstairs and got back in bed. Megan jumped up on the bed and looked at me as if saying, "What's going on? We are supposed to have breakfast now." I laughed and held her. I was looking out the skylight above my head. There was a small white cloud floating all alone. I said, "If you can do it, so can I." You look happy and comfortable. If I have to run this place alone I will! I got up and dressed, and Megan and I went down and had breakfast. I went to the window and looked out at the flower boxes and thought, I have to go down every morning and clip the dead flowers off. I have to check to see if they need water. I need to check the ropes at the end of the float. I need to check the firewood every week. I have to check the generator. Make sure I always have gas and oil for it, also for the boat. I will check the supplies like matches, batteries, spark plugs, food supply, dog food, paper goods. Keep the chimney clean, keep up on the fishing supplies, and check the water tank. The list was long, but the good thing was Rob and I did this together every time we went to town, so I knew what to do and what I needed. There was no one on my side of the island. There was a darling little house on the rocks next door. It belonged to Peter and Brunie. They would come in the summer and stay until August. Megan and I spent the day doing the flowers, chopping some wood, and checking all the things. We went to the water tank to make sure it was full. No big thing—if it's low you turn on the faucet. The thing to remember is to turn it off. I was a little excited about all of this. It was something to make my days go past.

When Rob came home he was helpful in some of this. Chopping wood and going in to the store with me. I loved driving the boat. When Megan and I are alone, we go out almost every day. I love to fish. Megan would dance around and make a little noise when I caught one. I had to teach her to sit and be still until I took the hook out. She was a very smart dog, and learned fast. I knew I had to take care of Megan and myself. If I am ill, I better take care, and rest and get well. I must stay out of the bush when bears or cougars are around, (hardly ever). I must take care not to let strangers who stop at the dock, know I am alone. One morning I was standing in the sun overlooking the vast skies and I was listening to the rise and fall of the water. It seemed to say; I am your strength, courage, passion, and your tranquility. I would do this alone and I would do just fine! The first thing I learned when I came here was to accept the fact that if you needed water, you had better find a way to get it. There is no room for I can't. As weeks became months I slipped into the rhythm of Island life. Deep within me there was a love of the island, the rich soil, and all the greens that grew there. It was 5:00 a.m., Tuesday morning, when I opened my eyes, looked at Megan and smiled. "Good morning Megan." There was not a blink to be seen. I pulled back the covers and crept out of bed. It was still very dark, and very cold. I stopped by the stove and added two logs, then made my way to the bathroom, and here came Megan right behind me. We did this everyday. She waited until I put the wood in, before getting out of bed. She and I have been here alone for 15 days now. Rob is working a double shift. We are staying in most of the day because of the ice. After breakfast, we'll go out and I'll chop some wood. I like to keep ahead of it so it has a few days to dry out. I chop wood while Megan looks around to see if there are any changes during the night. I carried five loads up two flights of stairs. It's all stacked against the wall unit Rob made.

Secluded Rendezvous

We went back down to the kitchen. I returned to the living room and lit a fire, then returned to the kitchen. I had my oatmeal and tea, and Megan had her breakfast. We went to sit by the fire in the front room. It was nice and warm. Megan got very excited and ran over to the front windows. She looked at me and barked. I went over to her and heard the boat. It was Ken and Marlene. They were pulling up to the float. I went out to the porch and warned them to watch out for the ice. Ken was carrying their little dog, to Megan's delight. They have such fun together. We had cookies and tea. They don't stay too long because time is money to them. They work very hard out there in the rain and cold. They usually do well fishing. I'm happy for them. Sometimes I hear the boat and run down to the float to see them waving and pulling away. I look down knowing there's a fish laying there for me. How nice! They yell, "Enjoy the fish we'll see you next week."

One day I was on the roof sweeping off the pine needles when Ken and Marlene stopped by. Ken yelled, "Where are you Carolyn?"

I shouted, "Up here." Ken looked up and yelled, "What in the ---- are you doing now? Get down from there."

I came down and explained what I was doing. Ken was angry. He said, "What if you fell, we wouldn't find you for a week, and what if you just broke your hip, but couldn't get up—you would die from a broken bone!" I saw how worried they were and I said, "I won't do it again." Did I? I'll never tell... I do know that he was perfectly right. The next day it rained so hard! Megan and I stayed all day. I made cookies in the morning. I washed windows inside, and built a fire. I sat in the window seat looking out onto Calm Channel. I counted at least 50 fishing boats speeding by,

Carolyn Begg

hands gripping the wheel. Some are anxious to beat the southeasterly winds. Some are in haste to get through the Arron rapids, which will reach their height in about an hour. The salmon run is on. Poor fishermen. They must risk their lives; they must fight for their existence. They must pray that they live in faith, if they live in fear they do not make it. And what about the fish? Do they not fight and risk their lives climbing and struggling to get up stream? Do they live in faith that the food will feed them? Do they not fear when they discover that the food has a hook beneath it? However, neither the man or the fish will survive without fight, risk, fear, faith, or prayer. Do fish pray? I was in bed early with Megan and my book. I always have a book to read. I finish one and start another. Rob and I are both readers. He has his Masters degree in Librarianship.

Chapter 23

The next day there was a loud blast on the horn from a ship. I went down the ramp and the Captain of the ship yelled, "Where do you want your washing machine? Where is your dock?"

I yelled back, "We don't have one yet, we were told you would deliver this summer."

Well, the crane came up and on it was a washing machine. He said, "We are putting it on here."

He began to lower it on the float. I asked if it would hold on the float. He said, "It is not my problem, it is yours now." The crew yelled at the Captain. "Let us take it up for her, or we can push the float and she can tie it to the ramp. "We're leaving," the Captain shouted and laughed. I could not believe this man. He thought the whole thing was a joke. It was like, you want to live out here lady? Then you deal with it. The ship was from Vancouver. As the ship pulled away, the crew men, looked at me sadly and waved. One man yelled, "Are you okay?" I yelled back, "Don't worry, I'll be fine thank you."

The wind was really blowing. The waves were causing the float to rock up and down, and the washing machine was sliding back and forth. It was hitting against the 2x4s that were nailed around the sides. I took Megan in the house. I went to the basement and got lots of rope. I went to the float and got on my hands and knees in front of the washing machine. I said, "Lord, should I try this? If this machine comes my way, and falls on me, what will happen to Megan? I need your help." I pushed the machine to the corner and tied it to both sides, bringing the rope across the top both ways. I started up to the house and I thought, I'm done now. Whatever happens is supposed to happen. I went inside, put Megan on my lap, sat by the fire, and we both fell asleep. We woke up and the wind had died down. The machine was still there.

We took a walk on the beach, came in and went upstairs taking our dinner with us. It was cold. I had a hot bath and sat in the tub with my book. I looked like a prune when I got out. Rob called and said he was coming home. He got the machine up, and said he would be back tomorrow. He said he had a big surprise and would be here early. Megan and I went to bed. In the morning I got dressed, tended the to the fires, put the oatmeal on, and fed Megan. We walked outside. I heard a strange sound of a boat. We went out front and looked down the channel and here came Robbie pulling a long, long, dock. It was at least 200 feet long. He brought it up to the raft. I said, "Robbie, it is wonderful! Where did it come from?" He said, "I bought it from a fish farm." Rob dropped some anchors. We went in and ate the oatmeal. We drove our boat to Middle Rendezvous. We had our tools, wood pre-cut and everything we needed to build a small dock. The dock was going to be 8x16 feet. This dock would attach to the ramp. The long dock was to be cut in half. One, 120 foot piece would come off this one. The other piece would be cut into four pieces, to use as fingers coming out from the side of the long dock. There would be spaces in between each one to allow boats to park. We made good progress with our dock. The tide was just right, allowing us the beach to work on. We tied it up at night, and would come back tomorrow at high tide to finish the top, and pull it home. We had a nice spaghetti dinner, lit a fire and went to bed.

Secluded Rendezvous

We were up and out to Middle Rendezvous. The tide would be just right at noon, to pull the dock home. This was so exciting. Imagine getting out of your boat, tying up and getting on your dock, and walking up the ramp. I couldn't wait for this. We had a couple of hours of work to do on it. When this was done, we brought the dock home. Rob jacked up the ramp end, that was attached to the float. We pulled the float up on our beach. We slid the 8x16 foot dock under where the ramp was to be attached, lowered the ramp down to it. Rob bolted the plate in place. With ropes we pulled the dock in place and attached it to the smaller dock. It was wonderful! We finally had a dock. We took a walk down it and sat in the middle of it. Rob explained where he wanted to put the fingers. He said he would cut them up tomorrow morning, they are small and it would go fast. I told him what I always tell myself, that is, you have the rest of your life. I should tell myself that more often. My dad used to say it to me. I didn't know what he was talking about until I grew up.

We took off the rest of the day and walked up to our little house. We loved it there. It brought back lots of good memories. We talked for an hour and walked down the road. It was a different life up here. The moss bluff, the picnic under the trees. Growing our vegetables. It was warmer too. It was windy and cold by the water and the boats were there. We loved both places. We walked back down and sat on the deck. We couldn't get over the ramp with this long dock attached in front of our lodge. What a good feeling. I asked how a salad and grilled cheese sandwich sounded for dinner. "Great," answered Rob. We had another early dinner and early to bed.

We were on the dock after breakfast. Rob cut the fingers out on the beach and we pulled them one at a time to where they would be attached. We had to replace some of the foam blocks under some. We went up on our front deck and looked down at the whole thing. It was fantastic! It looked so big spread out across the water. There was lots of room for boats. We knew the pilots would be happy. The beach was covered with rocks. There were big rocks, little rocks, and huge rocks. I said, to Rob, "If I were to pick up those rocks and throw them in the water, where they came from, I could make a beach. It's all sand except for those darn rocks."

Rob told me I couldn't do it, there are too many and they are too heavy, it would take you forever. I told him it was just an idea. I thought, I will do it! Rob had a few days left before going back to work. We took a few days off. We saw Ken and Marlene, went fishing, went to Heriot Bay, and he drove me around the island to take pictures. I got some real nice shots. I wanted to make cards to sell to the customers, when they came, in the summer. We had a nice time together this "time off." It was time for him to get back to work.

One morning, just after he left, I decided to make a beach. Megan and I went down to the beach and I began to throw the rocks out into the water. I had to put them far out and not all in the same spot, or the boats would hit them. This wasn't a beach boats came in to, but I didn't want to take chances. We only came in here when we wanted to work on the beach. I did about forty rocks and I was tired. Some of them were big. The next day I decided to only do 20 a day. My back was hurting so I went to ten. My back kept hurting and I kept throwing. I made a beach. I ran and got a beach towel and Megan and I had an afternoon on the beach. It was so fun. When Rob came home, I went down to meet him. He told me I was walking funny, was there something wrong with my back? I have had two slipped disks for

Secluded Rendezvous

years, so I just said, "Not much." I walked him to the side of our deck. He looked down, and said, "Wow, you made a beach. I knew as soon as I saw you what you had done—it looks great."

I said, "The next time you come home, you'll see a pond there by the waters edge."

He asked, "What for?"

I told him I was going to make one out of the rocks I kept aside. It would be a large circle with a fish net over it to keep the birds out. We could bring some oysters here, just enough for dinner when customers are coming.

He helped me build it. It was fun. We went in and I said, "How about taking a trip to mom and Haakon?"

Rob answered, "Let's do it the next time I come home." I said, "Great." I asked him what he was planning to do, I knew there was something on his mind. He told me we needed a generator shed. I asked if we had enough wood. We went to check and we did. The shed would be 6x8 feet. Rob said, "We are going to sound proof it as much as we can so we will have a low ceiling. We won't have to stand up on both sides, just in the middle for service." The sides would have wood, and dirt on top of the wood. There would be large rocks in the dirt. We built this, and it turned out well. In fact, it looked real cute sitting out in the back. Of course, Rob picked the spot. I planted some flowers in front of it with flowers in between the rocks. We later bought a 10 K Generator. It was large enough to run the whole Lodge. It took us several days to build this. We finished and Rob had a couple of days left before he had to return to work.

We wanted to go to Heriot Bay, and walk around the shopping mall and have lunch at the Inn. We had a nice day, went to the little grocery store. On our way out a man approached us. He asked if we owned the new lodge on rendezvous island. We said we did. He got very excited and said he had a place on Quadra and would like to come out to see ours one day. Rob said sure. He told him he would be gone to work in three days. The man, who's first name was Glen, said he would come tomorrow. While Rob and I were having lunch, I told him I didn't care for this man. I didn't understand why he was so excited. Rob said maybe he wants to stay there and fish. I said maybe. He came and asked how it was in the summer. We said we were opening soon and had a few customers. He said it was so beautiful, and it should be full. Rob told him we didn't have the money to market it the way we wanted to. But we will soon. Glen said, "I have a friend who loves to fish, and money is nothing to him. Could I bring him out to see this lodge?"

"Sure," we said. When he left I asked Rob, "Why does he want to bring that man here?"

Rob said, "Maybe they want to buy in."

I asked, "What would that mean?"

Rob said, "To be partners." I said, "We wouldn't want partners would we?"

Rob said, "Sure, maybe they could get it marketed."

I said, "I don't want partners."

"Let's wait and see." Rob said.

Rob was home and they came out to see us. I heard one say, "God, this place is beautiful." The new guy's name was Gary. Gary came over to me and said, "Glen tells me you're having trouble marketing this place." I said, "It's not trouble it will just take a little longer than we planned. Everything in Canada is two and three times higher than home."

Gary said, "Do you believe God sends people to help you?"

Secluded Rendezvous

I said, "Yes I do." They left and Glen said he would be calling Rob. Rob and I never argued. We started back and forth. He wanted this so bad and was tired of waiting. I understood. I told him, it was ours, why do we need other people? We did this all ourselves. I always had heard it wasn't good to have partners. We went on and on about this.

He left for work. I held Megan and cried. Two days later they called him, he said yes to their deal, and we went to sign papers. I did not trust them. Anyway, now we were four partners. We all had a company attorney that Glen knew in Campbell River. I went on working with everything I could get my hands on. I was not happy. Rob was at the farm, working. He had been there two years now. He was leaving his job and coming home. He would work another double shift and be home in a month. I took everything out of the pantry and washed the shelves, and everything in them, and put it all back. I went outside with Megan to the side deck. Rob had talked about putting up a smoke house for fish. He said we would need to build a fire pit just off the deck and have a long pipe going up the slope to the smokehouse. I decided to start gathering rocks for the pit. I would build it. It shouldn't be hard to do. Megan and I went down to the beach and I got rocks leftover from the pond. I carried these up to the deck. I ran out of rocks, so I took some from Peter's beach. I knew he wouldn't care. He would be happy for me to make him a beach too. I went in with Megan, when I had enough rocks. We had dinner and we went upstairs. Megan and I got in bed early. I was tired.

Chapter 24

We were asleep, and I heard a terrible crash. Megan let out a bark, and I whispered, "Megan, no speak." We went to the window, looking down, I saw a large fishing boat. It had rammed into our ramp. I stood very still and watched a man get out of the boat. He stood there, looking at the house. Rob had our boat, and I had pulled the skiff up on the float in the bad weather. It looked as though no one was home. This man started to come up the ramp three times and would turn back. He never uttered a word, and he looked drunk. I picked the Radio Phone, and called the operator. I asked her to call the neighbors on the other side. No one answered. She called me back and said, "I'm calling the Coast Guard. Stay on the line." The Coast Guard came on the line and asked me what he was doing. I answered, "He went back in the boat."

They said, "We are at Big Bay Marina and will be there in five minutes, we are on our way." They told me to stay where I was by the window, fling it open hard and yell, "What is your name, and the name of your boat, and the Coast Guard is on their way they will be here in two minutes." I told the Coast Guard I would.

I also said, "I'm not afraid, I have a hunting license and a shotgun. I will shoot him in the legs, if I have too."

The Coast Guard answered, "Be sure he's in your house." Out here, we all had a boat phone, they were usually left on all night. If we were inside, we took the phones with us. One channel was used for the Coastguard only. Any channel could hear what was

Carolyn Begg

being said. I heard the Coastguard come in and tell him to stay in his boat, with the phone left on. They arrived in exactly five minutes. They got in his boat with guns and searched him and his boat. It was a rare thing for a fishing boat to do this. They are excellent seamen. The men took him out on the dock and searched him. They questioned him and told him to get back in his boat. They walked up the ramp to my door. They told me he said he had trouble backing up. I asked why he would be backing up going down the channel. He did have two little boys with him. He wants to sleep all night here. They said they didn't want him driving again with the boys. Maybe he needs to sleep. We think he's okay, just not to bright. He rented this boat. I said he could stay. They told me he has to be gone at daybreak. We told him that and he has agreed. He said he was taking the kids to their mother. "We will alert the Coastguard from here to her home in Vancouver. We'll be watching him along the way. We're wondering why he has rented such a big boat when he can't drive it, and how he got this far. Maybe he was running away with them. We don't know yet. We will find out on the other end. Carolyn, you're a brave young Lady to be here alone. Not a good idea. Everyone who had their radio on knows you're alone." I told them I would call Rob when they left and tell him I'm sure glad he will be here tonight to stay. They will all hear. I'll be fine. They turned to leave and said, "Carolyn, if you shoot someone that is after you, don't shoot them in the leg. Make sure they are dead. And, if they are on your porch drag them in the house." I thanked them, and they left. They said, "Leave your radio phone on."

Megan and I got up in the morning, I went to the window, looked out, and there he was fishing off the dock with the two boys.

Secluded Rendezvous

I walked down the ramp. I looked at him and told them, "I'm sorry, boys, but I have to do this. The sun had been up for three hours, and you're still here. I'm going in, you will see me in my window with the phone. If you have not left in five minutes the Coastguard will be back here." He got in his boat, with the boys, and left. I felt bad for the children, but I knew they could fish all the way home. I was happy he was gone. Three boats came up to the dock that morning, checking to see if I was okay. They heard us on the phone. How very nice of them. I felt very safe after that! Megan and I went down to the beach. I collected more rocks and stacked them. I got a bag of cement and the wheelbarrow to mix the cement in. I made a batch of mortar. I drew a circle on the ground to make sure a piece of wood would fit. I measured with the wood we cut for the small stove upstairs. I added inches on both sides. I decided on a good size and the circle was fine. This would be so fun! Like doing an art project. I finished by early afternoon. I wasn't sure about the height. I left it a little short. I could always add rocks after Rob checked it. I did leave a square hole to fit the logs in. I needed flat rocks for around the top. I knew these were on the upper road by the house, if need be. I will know more, later. Rob and I would get them.

I looked at Megan and said, "Want to go fishing?" She ran down the ramp to the dock and sat by the boat looking at me. I got in the boat and said, "Okay Megan." She made a jump in the boat and sat down. She was such a good dog. We took a long ride across the other side and trolled along the side for a Salmon. I caught a nice big one. Maybe thirteen pounds.

I drove around to Ken and Marlene's. I went very slow so they would hear the motor on the boat. Ken came out on the porch. I held up my fish. He called to Marlene, she came out, and he yelled, "Look, what Carolyn caught, and didn't ask me to go." Marlene laughed and asked me to come in. I told them I had to go clean my fish, unless Ken wants to do it. I drove away laughing. They are so nice. I took my knife out and put the fish on the cutting board Rob had built on the dock. I cleaned the fish and we went to the kitchen. It went into the freezer. I would keep it for Rob to put on the grill outside. The sun was hot. Megan and I sat in the shade on the front deck. I brought out my book and read for a couple of hours. We went in for dinner. Then we were upstairs for the rest of the night. In the morning, I chopped a stack of wood. I like to keep ahead of it. I heard a boat and saw Peter and Edde's boat. Edde waved and pointed at me to Peter. He shouted from the boat they wanted to tell me they were here for the summer. They won't come now, but will come after they're settled. They would call. They were fighting the waves they made so they backed out and left. It was so good to see them. They were a month early this summer. It's so nice when everyone comes for the summer. Boats are coming by, people are hollering and waving. The Island comes alive. I'll have to get supplies ready for the summer parties. Fun! I finished the pit. I believe it is the right height and I wanted to clean everything up. I went down to the beach and got some shells I had put aside while working on the rocks.

Megan and I were in the house when Peter called and said Edde wanted me to come to lunch tomorrow. He offered to pick me up. I told him I would drive over, but thank you. He said, "Edde said to bring Megan."

We went at noon the next day. Edde looked as though she had everything put away and together. She is very organized and a hard worker. We had lunch on the deck. It was a lovely spot, overlooking the South side of

Secluded Rendezvous

the water. They have a large beautiful new boat. I had a nice day with them. She loved Megan. We walked over their land and they told me of their plans. A good friend and his wife were building a house in the apple orchard. These people were also from Germany. They would be here next week. They were all going to be together every summer. They were good people. Peter and Edde had a son and daughter who would be coming in the summer also. It would be a busy, fun summer for them. Megan and I got home in the late afternoon. We fished on the way. I caught five cod fish. This would be great for our dinner when Rob came home. I tied up the boat, cleaned the fish, and was putting them in the refrigerator, when the phone rang. It was Rob. He would be here in three weeks. He was happy. I was, too. I had a sandwich, and Megan wanted dry dog food. That's all I give her. She's not allowed to have people food.

Chapter 25

We were in bed early, as usual. I woke up in the middle of the night. Something was walking on the roof, above the bed. I reached for my gun under the side of the bed. I laid it across me and woke up Megan. I said, "It's okay, Megan, no speak." I could hear the roof cracking. I was thinking this thing is big. It is either a man or a bear. I reached for the flashlight and turned it up toward the skylight. If it was a man, I could see him, but he couldn't see me. I brought the light down to the gun. I heard the noise stop, the thing was walking away and I didn't hear it again. I was awake most of the night. In the morning I called Peter. He came right over. He, his son, and I went outside. He was walking all around looking for prints. At the end of our roof, there was a place where the bank from below it, nearly touched the corner of the roof. We could step up on the roof from there. This area was covered with tall winter grass. I noticed the grass was flat like something rubbed it to the roof. I called to Peter. He came and looked hard it. Then he said, "It's nothing Caroline, just some birds playing." Peter was a big game hunter in Germany. Edde just told me yesterday about all his trophies. I thought what's wrong with this man? Birds don't leave a trail like that. He said, "I didn't find anything, no tracks Caroline, it was nothing." He started walking away and down to his boat. I thought he is acting funny. Maybe he thinks I'm just afraid.

When he got in his boat I said, "Thank you, I appreciate your coming." He looked at me and said, "Caroline if you're so frightened, don't go into the woods, don't even go check your water tank," then he laughed and drove away.

I didn't know if he was making fun of me or not. So I yelled, "Peter, I'm not afraid there was something on my roof!" I did what he asked, and stayed out of the woods, but I didn't know why. Peter was from another country and we don't know each other yet. I tried to forget about it, but still I knew there was something. I also knew deep down, that Peter knew what it was. While Megan and I were quietly eating our dinner, the next day, I heard one gun shot. I knew it was Peter. When I saw them the next time' I asked if anyone heard the shot they all said, no. However, a year later Peter showed me a chain he wore around his neck. It had a Cougar tooth on it. Edde told me he shot Caroline's Cougar. He didn't want me to be afraid to walk in the woods. It was a rare thing for a Cougar to be on the island, and he knew how I loved walking in the woods. Megan and I were sitting on the deck. The white caps were building small, but many, pushing to show their powerful strength. After hours of using their extreme pressure, they release their forcefulness and resume to rhythmical song until the sea is once again a body of peacefulness, calm, smooth with continual flowing. Again I say, "How can one not believe in God, he's right here now!"

The phone rang this night. It was Randy. He wanted to come next week, and bring two of his friends, Gary and Pete. They wanted to stay with me for a week. I told him they could stay as long as they wanted. He would set the time with them and call me back. The next morning he called and said they would be here Sunday. I told him where to park the car, walk down the road straight to the dock. I would be there. I was so excited I jumped in our boat the next morning, and went to Ken and Marlene. I forgot they were fishing today. They would be by here tomorrow to check on me. I would tell them then. Megan and I came back home. I called Rob and he was happy for me. He would miss them by two weeks. Ken and Marlene came the next day. They had lemonade and cookies. I told them Peter and Edde were here for the summer. I also told them about the noise on the roof. Ken said that was odd for Peter to act that way. I said I guess he thought I imagined it. He doesn't really know me yet. Anyway, whatever it was, it didn't come back. I showed them my fire pit. I asked Ken if he thought it would work, He said, "Sure it will. You can come build me one." We laughed. I told them Randy was coming. They said, be sure and bring them over, we want to see Randy. We will all have a beer together. It was time for them to leave. We said goodbye. I told them how much I appreciated their coming here every week and I'd see them again. Marlene said next week, if not before.

The next day Megan and I went to Heriot Bay. I docked the boat and we went to the store. I needed food for the boys. I tied Megan outside, and went in and did my shopping. The lady in the store said most of the dogs bark and Megan never did. I bought her some doggie treats. We got in the boat and I drove over to the gas pumps and filled the tank and our gas cans. I knew the boys would be going fishing. Megan and I got back home safe and sound. We had a great ride back. The water was perfect. We passed Peter and Edde on the way. We were going home and they were going into Heriot Bay. Seemed funny to pass someone you know, the horn beeps, and we all wave. It's like that in the summer. In the winter, most of the time you're all alone out there. You can go to Heriot Bay, and back and never see another boat pass. Both ways are great with me, just different. I love driving the boat. Out here you don't have to look for traffic, and be with the people on the road who don't know how to drive. I docked the boat and went in to the kitchen to put the food away. I told Megan I was going to buy her a pack-sack. I had made three trips back and forth, from the kitchen to the boat.

Chapter 26

Two days later we went back to Heriot Bay. There on the dock was my son, Pete, and Gary. They were all waving. What a wonderful sight. I was so thrilled, I couldn't get out of the boat fast enough. It was so good to see all three of them. When we got through hugging and with the tears, we packed the boat. They had a very light load. This was good. I was in a 16 foot skiff with a small motor. After the boat was packed we headed for rendezvous island. The water was smooth. I had Randy sit with me in the back. Pete was in the middle and Gary in front. The balance was good. We drove along, Randy and I talking. Pete turned around a few times to ask a few questions. Gary was quiet. Pretty soon he turned around and said, "How far is this place?" I answered. Then every so often came, "Where are we? Are we getting close, there's no one around here, where are the other boats?" I was laughing inside. However, I knew all of his questions made sense, from a young guy from the city. He even asked if we were lost.

Finally we saw Rendezvous. I got to say, "See that island in front of us? That is Rendezvous. We are almost there." We went around to our side. We pulled up to the dock. Gary said, "My God, this is huge, I thought we were going to some little cabin." He felt much better. We brought the bags up, I showed them their rooms.

We all took a walk up to the house. Randy had not seen it finished and he really liked it. "Do you miss living up here," he asked. Sometimes honey, I love it here but I am also close to the water in my heart. He said he was happy for me because we had them both. We went on a long walk. They wanted to see it all—and I wanted them to. We walked down the road past Judith. I told them about their home and how they had built. We passed a little house made by a guy named Berry. He lived alone and had built his own place. We walked by John and Carol's cabin. Then we walked up the road to Sam. He was not there; I knew because his boat was not docked in marks bay. I wanted to show them the house he built and his lovely garden. It was a cute little place and he did it alone. Hard to do when you don't have someone to hold the other side of the wall while you nail.

    We turned around and went back home. Every one was tired. We sat out on the deck. Megan fell asleep on Randy's shoe. Randy loves animals-like his mom. Randy told me about their drive to Canada. They had a fun time. They are good friends and tease and laugh a lot. We went in and had dinner. After dinner I asked Randy to fix them some drinks. I took my tea-cup and sat in the window seat. It was sixteen feet long and three feet out on each side. The guys wanted to know how we did this, and how long it took to do that. They wanted to know about the windows we were sitting in front of, they were very high. I told them the story. We had a ladder outside and one inside. Rob was outside. We wrapped a blanket around the window and Rob leaned the ladder out on a slant. He put the window, leaning it against the blanket, pushing it up as he walked up. I was on the inside ladder hanging out the hole, reached the top of the window and pulled it up and held on until Rob came up my ladder. We were both pulling up until we ran out of elbowroom, and I went down the ladder. I went up the outside ladder, held on to the bottom of the window, we moved it in place and Rob nailed it in. Then we did the other three. They were much easier, because they could be nailed from inside. The ladder outside was not tall enough to reach, for the high window. They were all hard because of their weight. Each time we did this I said, "Lord, please don't let me drop this." I was always happy when Rob climbed down the outside ladder. We only had two very high windows, and I was thankful. I told Rob, "I can't believe we did this," He said, "I can't either." It's always a wonderful challenge and so rewarding when it's done.

"You guys must be tired by now, let's go to bed." And we did. I got up early and was cooking a big breakfast. Randy came in, hugged me and said, "Thank you mom for letting us come." I told him this was his place too, and he could come anytime.

We thought Gary and Pete were still in bed. Randy and I took our coffee in to the living-room. There was Gary in his socks, in the window seat. He looked up and said, "I don't think I can stay longer than two days." We didn't answer. Pete came in and said, "I looked out my window, it's unbelievable out there." We went into the dining room and had a big breakfast. We talked about what they were doing at home. When we finished Gary came in the kitchen and I was washing the dishes, he asked if he could help. I knew he wanted to talk. I gave him a towel. He said, "Carolyn how can you stay here twenty days a month all alone?"

I answered, "It's usually only ten, twenty days is a double shift." He told me he would go crazy. I asked him if he believed in God. He told me he did. I told him a few stories.

"One time I was working, digging out cement from a container that was on the beach for one year sealed. I suddenly became very ill. I doubled over in pain, and I was sick to my stomach. Rob helped me upstairs to bed. He called a man on the island who was a doctor, visiting, he said it was something I ate. I told Rob I was really very ill. I was passing blood from my stomach and bowel. I prayed all night. I asked Jesus to hold my hand. I knew I was in trouble. The planes can't fly out here in the night. Rob called the doctor, in the morning. He didn't want me to fly, because of the altitude. They flew me out very low over the top of the water. When I got to the hospital, they had trouble helping me, because my veins had collapsed. Rob had to take me to the hospital in Vancouver. By the time I got there I was dehydrated. After it was over they told me I could have died. The container I was working out of on the beach was contaminated. When the doctor said I couldn't fly, I prayed. When the guy on the Ferry said they were full, Rob ran in the office, they didn't believe the "Emergency" story he told them. They took one look at me lying on the seat of the truck, and said they were making room for the truck for us, and we would get on. When any of these obstacles came, I prayed. I asked if he wanted another story, and he said yes.

One time I went to where Rob was working, and had dinner with him. He misread the tide chart book. He told me when to leave to get through the whirlpools. I left, drove around the bend, and right into them. They were going full strength. Once you are heading into them, you can not turn around or you will get pulled under. So I kept driving through them. I said, "God, please help me, I've never done this before. Please take me through this mess." I was driving in-between huge circles with very deep holes. We got through. Gary, I know I felt his presence. The point to these stories is, I am never alone. God and Jesus are always with me. I'm never afraid. I've been in storms so bad and had to pull the boat out of the water on to the dock, sliding all over the place. I shouldn't have walked on the deck,

Secluded Rendezvous

let alone, pull a boat out of the water. As I was pulling, I asked for help and the boat came up. I could write a book on these stories alone."

He said, "I wish I had your faith."

"Gary I don't just talk to them when I need them for help. Sometimes I say, 'I know you're here. Thank you for this beautiful day, or thank you for bringing these Whales, that are swimming by.'"

Gary said, "Whales?" I told him I have my own Marine World right here. Randy and Pete came in and asked what we were doing. I said I was talking and Gary was listening.

They told Gary the fish were waiting. Before they left Gary told Randy, "Your Mom told me some wonderful stories. I'll tell them to you in the boat, if it is okay, Carolyn." "You bet Gary. Here, I made you guys lunch this morning." I showed Randy the chart, and told him where the whirlpools were. He said they would take a little tour, and fish right out in front. "Tonight we'll study the chart and tide book. It's okay to drive around the Island. The fish run out in front too." They spent half the day in the boat, and came back with fish. We had fish for dinner. I had the rice, vegetables, and salad ready—I knew they would catch fish. At dinner they told stories on each other, who lost hooks, and who lost the fish. We had a fun time. I went upstairs early and let them talk and laugh together.

On the third day when I came down, Gary was sitting in the window seat with his book. He looked at me and said he wasn't ready to go home. He liked it here. I said, "Good, you'll all stay with me." I went to the kitchen, put on the coffee and sliced some melon. Randy came with my hug, and started setting the table. "I'm so glad to be here mom, it's nice to see how you live. I do worry about you all alone, I know what you told Gary, and I know you have always had that faith and taught it to me, but you work so hard here, mom." The boys came in for breakfast. We ate the breakfast, and went out on the deck. I took them to my beach. Megan ran ahead and stood at the oyster bed and looked at Randy, as if to say look at this. We all had a good laugh, and Megan was such a cutie. There was a big rock sitting there. Randy said he was surprised I didn't move it. I said I tried and it would not move. They asked what I used. I went up and got the peavey. Randy asked if he could try. He tried and moved the rock. I was so happy. I told him every time I looked at that thing I wanted to try again. I said, "There are some things I know when to leave alone."

Gary said, "I do not believe that."

The guys went on a hike. I told the them there were no poisonous snakes or poison oak. There are some ticks, but we have never had any on us. They hiked around a few hours and came home. I had the feeling that they didn't want to leave me alone for long. I really appreciated this because I wanted them to have their time, but was waiting for them to return. I tended to my flowers: cut off the dead ones, fluffed up the dirt, and fed the flowers. Megan slept in the sun. The boys came back, and told me their adventure. They thought it was all so beautiful. They asked it they could chop some wood. I gave them a lesson. I don't know if they needed it, but thought it would be a reminder. I know they haven't been around wood for some time. They chopped a big pile for me and I was pleased. They asked where the water came from. I told them the story, and said I never got back there to make it bigger, and I want to do this, one day. They wanted to walk the line to see it, so we took a walk. Randy went and got a shovel. I asked what that was for. He laughed and said,

"Maybe I'm going to dig." We arrived at the end of the line, which was in the hole. I pulled the weeds around the hole and Randy dug it bigger. It was great! We left the shovel and walked up the road. We met Pete and Gary. I took them to the spot where I used to sit when we lived in the little house. They loved it. We picked up the shovel and the guys admired the digging Randy did. On our way back we were all quiet. Pete told me it was so nice of me to share that special place with them, and thank you. I told him that I was so pleased that my son had such nice friends, and they could come back any time and chop wood. They thought that was pretty funny. It was late in the afternoon. I asked if they wanted to see if Ken and Marlene were home. They said they did so we all took a boat ride.

Ken and Marlene had just gotten home from fishing. The guys said they saw Ken's boat out in the water, but kept going. They didn't want to get tangled in their lines. Ken said, "Thank you, people forget about them, and drive right up. It can be a mess."

We all walked to their house. We spent the rest of the day there. The guys had a few beers, told stories, and we left for home. None of us wanted a big dinner so we all made our own sandwiches. They laughed when Pete said he was ready for bed. He said, "I can't believe I'm going to bed this early." Randy and I stayed up a couple of hours. He came upstairs with Megan and me. We lit a small fire, the nights were a little cold. We had a talk about our personal lives and then he went down to bed. I am so blessed to have such a wonderful son and I am so very proud of him. He is a fine person. The next day was the last day for fishing. It was Thursday and they were leaving Friday. I won't think about it now. How very sad for me, but they do have to get on with their lives. They were all up when I came down. Gary was sitting in his usual place. I heard him say he could stay here. He didn't want to leave. Neither did Randy. I'm not sure about Pete. It was time for them all to get back home and their jobs.

We had breakfast and I packed food and water for their day. I got busy with "nothing things" to pass the time. They were only gone a few hours. They took a ride to big bay. I had told them the time to go through the whirlpools. I explained how they worked one night. They had a good fishing day. They cleaned the fish and brought them up. I had brought up oysters and clams. We had a very big dinner. I was happy they seemed to have had a good time. I said, "I will clean up and you guys go out on the deck. I'll meet you there." They would not stand for me doing this alone. We all worked together, and were done in no time. It was fun, too. It was a lovely evening. Megan took turns sleeping on the guys' shoes. We sat and looked at the stars. The stars here can be seen more than one can imagine. The whole sky lights up. It is spectacular. They couldn't believe it the first night. Nor could they believe the phosphorus in the water. When you splashed the water at night, sparks fly everywhere. When you drive the boat at night, the back of the boat has sparks following you. It is so beautiful and fun, it takes your breath away. Well, we waited all we could, to say good-night this night. They were leaving at eight in the morning. Their drive was a long one so we all said, "See you in the morning." I said I would wake them up at six. We would leave at seven thirty so they can catch the ferry. They said they would only have toast and coffee. Randy came up when I was in bed. We hugged and kissed, and cried, we both said I love you and will miss you.

In the morning, we had our light breakfast and went to the boat. It was hard for all of us. I drove so

Secluded Rendezvous

they could see the view. I hope they return. Megan was very quiet sitting next to Randy. She was enjoying the ride. She loves the wind in her face as much as I do. We all hugged and had tears in our eyes and tried to say goodbye as best we could. We were trying to be brave- it was hard. I left them on the dock and drove away. When I got out a little ways, I waited out in the water to see the car drive down the hill and on the road to the Ferry. I cried most of the way home. We were leaving each other again.

Chapter 27

The next week went fast. Rob was leaving his job in two weeks. I had some customers. The customers are always nice people. None of them liked the idea that I was running the lodge alone. I was okay with it. It was company and fun. It was too much work for one person, but I didn't do it every day. One time I had a man come for three nights. There were no other customers. He liked his liquor. He wanted me to have dinner with him. I told him I didn't eat with the guests. He was mad. I sat down with him and my cup of tea. He started to get very friendly and drunk. I went upstairs and called Rob. He said he would come and stay overnight. He stayed until the man left. That was the only time I didn't enjoy a guest.

In the morning, Megan and I were sitting on the deck and I heard a terrible cry, it was a cry I had never heard before. I looked toward the sound on the beach next door and saw something there. I ran over to it and there was a mother seal giving birth to a baby seal. I was so excited. It had just been born. It was darling. The mother seal looked at me and ran for the beach. I felt so bad I left. The baby seemed to be okay. I ran in and got my camera. Mama sat watching me as I took my quick shot, and went back to our deck to Megan. Megan and I watched from the deck and mama seal returned to the baby. It was a wonderful thing to see. Poor little Megan, she paced back and forth, she was upset about all of this. I heard a boat pulling up at the dock. It was Peter and Edde, and Ebb and Anna. Ebb and Anne had a house built in the apple orchard next to Peter and Edde.

Ebb was a Doctor in Germany and Anna was a retired ballerina. She danced at the Metropolitan of Arts, in Europe. She was a lovely lady. She was in her seventies and still had the body of a young ballerina. It was so good to see them. Of course I told them about the seals. We were all excited! They were very good to me. Ebb and Peter will go by the lodge and blow their horn on the boat. If I don't come out on the porch, they come on the deck to see if I'm okay. The four of them came up and we all sat and had a chat. It's always nice to hear about their life in Germany, and also on the island. We are all on a different journey. They came to ask me to lunch the next day. We all went to big bay and had a wonderful time as we always did. They dropped me off on the deck.

I ran up to get Megan, she was so happy to see me. We sat on the deck and I was watching the water. The white caps were riding high on the waves, crashing, then folding down into the troughs and through rising, swelling, then down to their roll. It repeats and repeats and suddenly the sea is once again calm, drifting slowly to its destiny. And down the channel, here came Robbie. Megan ran down the ramp, her ears flapping in the wind. Rob picked her up in one hand, duffel bag in the other, and they came up the ramp. I was so glad he was a part of this place again. It sure took the pressure off me when he was here. I missed him so. However, I was brought up by my father to know that a person can do what they want to do, and no one can or should stop them. Rob has always done what he wants... and I let him. He worked there for two and a half years. That was a long time for me to be here alone. He was here every ten days, so I wasn't completely with out him for two and a half years. But by the time he got home, it seemed it was time for him to leave again. Well, that was yesterday. I said, "Hi honey, it's good to have you home." I was really happy. We hugged and kissed. I told him I missed him, he said he missed me too. He put his things away, and I fixed dinner. We ate outside. We had a nice dinner and talked for hours. He talked about his work and I talked about this place. We stayed up to see the stars and went to bed early. Megan got to have her own comforter on the floor next to the bed. She was happy, and so was I. I had worried about taking her off the bed. She didn't seem to care.

We were up early, having breakfast, when the phone rang. It was Glen and Gary, the partners. The season was really just about to begin. They said they were coming out to see the place and talk. Then came around noon. The two of them walked around the inside looking at everything. They came over to us and said, "We need to get new carpets,"

We said, "These carpets are still new. Very few people have walked on them."

One of them said, "All the furniture here has to go. This place has to be first class. We are getting Berber carpets, and new furniture in every room."

I asked Glen what was wrong with the furniture. He repeated that everything had to be first class. They stayed overnight and assured us that it would be better to have everything new. They left the next day. I was not happy about any of this. I told Rob they have taken over our Lodge. It will not have the charm that this place has. Everyone tells us how warm and lovely it is, it will all be gone, I cried.

The new furniture started arriving. There was a green leather couch, and loveseat. There was a pool table, dinning room tables and chairs, sleigh beds and an armoire for each bedroom. There was a three thousand dollar orca to hang, on the wall. Five new complete bathrooms arrived with new showers and sinks. A new linen room was built, and a new bedroom. All my curtains and bedspreads that I had made were taken out. They said, I was so good at decorating I could make new ones. The living room did look nice, like an office. The charm was gone, the place looked like a Motel. They bought bicycles and helmets, very expensive ones. These were to be ridden on a dirt road. You had to walk up a 250 ft. bluff to get to a logging road. It was full of rocks and holes. I did not understand any of this. Robbie and I were arguing all the time. They even bought kayaks to paddle. These were to be used in the water with the boats and ships. Glen and Gary would hold meetings. At these meetings I would ask them why they were spending all this money. They would answer me by telling me it has to be "First class, Carolyn!" I knew something was wrong, but I didn't know what.

The lodge went on the Internet. Fishing guides arrived. These guides all knew us. They watched us as we built our lodge. They had been going by with their customers over the years. Our customers started flying in. When we signed the papers with the partners, we had a meeting. Glen said, "I'll get the Guides, and take care of the bookings, Gary will take care of the Advertising, and Rob and Carolyn, since you two know the area and built the Lodge, you take care of the customers." We all agreed. Rob and I would get a call saying on this date these people would arrive at this time. It all seemed to work. I was unhappy. The Lodge, ten acres, and our little house, no longer belonged to us. No one knew I was unhappy except Rob. He was happy with the "new look," and all the people coming to stay. I understood this, and I was happy with all the customers, and it was fun. However, I could not get over my women's "intuition". I knew something was wrong.

Secluded Rendezvous

Gary was the one paying for all this. Glen was buying and Gary was paying. Why? I did enjoy being able to take care of the guests. They were what this place was all about, and they were paying a very good price. I gave them what they deserved. They were all nice hard working people that were on vacation, away from their problems at work or home. I knew because I had been there, too.

One time, when Glen was there, he said I was giving too much to the customers. I told him this end of it was my job, and if it were not for the customers, he wouldn't be here. The customers had a wonderful time. At night they partied, played cards, pool, or sat by the fire and exchanged stories. I got up at five every morning and I started baking bread, cookies, and making breakfast. I put all the lunch makings on the counter for the fishermen to make their lunch. After I fed them, they were in the boats, and on their way. Some of them fished on their own and some went with a guide. After they left I cleaned everything and went outside to tend to the flowers. It was a nice relaxing time. I would go back in and cook for the sleepyheads. These people came out a few at a time. There were usually several women who didn't care about fishing. They were happy to sit in the hot tub, or just on a chair on the deck, and get to know each other. Others went for a walk, or would sit and read. This was my time to clean bathrooms and tidy things up. I started dinner early. We had five rented rooms and a room for guides. Most of them preferred to sleep in their boats. Each room had a two person occupancy. We seldom had children, but had cots to bring in if need be. We usually had from eighteen to twenty people for dinner. I prepared all that I could in the late afternoon. The lunches were easy. Someone would stroll in really wanting to help. I loved it, we would talk about their lives and ours. Rob was always fixing something, taking care of things, chopping wood, cleaning the hot tub, or whatever needed to be done. After the men came in from fishing he would clean the boats, and fill the boats with gas. I always had dinner at six. The men were on time. I never had one late, and I appreciated it. Almost every time a man would come in with a pot and say, could you cook these for us? There would be oysters, clams, or mussels. I always had a big pot of water on the stove, ready for them. Dinner was fun! One of the guides used to stand up at the table, lift his wine glass, and yell, "Let's hear it for the cook." They all raised their glass and yelled, "Hear! Hear! Hear!" I loved it. This guide was Larry. He would walk in the lodge, after fishing, and yell, "Honey I'm home." He made all of us laugh. He was nice, and a fun guy.

One night we had a luau. Rob made a table, large and low, to use on the deck. I had Rob cut some branches to cover the table. We used these branches, along with some flowers. We had Hawaiian music, and I did the Hula. I was pretty "rusty," but I got through it. It was fun we all had a good time. We always invited the neighbors on the island. It was good to all get together. Everyone drank too much, and we all got silly, it was good for all of us. The next morning we all had an extra hour of sleep. The fisherman slept in. I couldn't believe it when they were not up at five. I gave them a bad time telling they I got up for them and no one showed up in my kitchen. They were a fun group.

One morning Peter and Brunie drove up to our dock. They had just come from Vancouver. They lived in Spain, and spent a few days in Vancouver before coming to the Island. They lived next door to us. Rob helped Peter get his ramp in the water and everything in order. They are nice people. It will be good to have them here again.

Secluded Rendezvous

The partners came out with their wives and children. There were ten of them altogether. One of the partners announced that they all were partners in the lodge. One of the guests looked at me, then at Rob, and I said, "Yes they are." When I was in the kitchen cleaning the guest at the table came in and asked, are those two couples in with you and Rob? I answered yes we are all equal partners. He said, "Then why are you waiting on them, and their families, they act like they are guests." I replied, "I'm sorry, I can't talk about this, but I know what your saying is true—and thank you." This group of people left, their vacation had ended. It is hard to live with people for a week, night and day and say goodbye. Some of them sent us gifts, recipes, and letters. How nice. A group of very successful businessmen, came to a different lodge every year. This year our guides brought them to us. They were a wonderful group of men. We were told they played cards, stayed up late, drank, and don't be shocked by their language. They stayed with us for ten days. I loved them all. They played cards until all hours. We stayed up with them. They went out on the porch to smoke. I had tins out there with sand in them. Once in a while I heard a slip of a word and it always followed with, excuse me. They were polite and appreciated everything that was done for them. A couple of them came in the kitchen and asked if they could help. One man didn't ask, he picked up a towel and said, "I'm drying. I want to hear about you two. We can't get over you two building this place by yourselves. Where are you from, where did you work, how did you find this place?" There were a lot of questions. He was genuinely interested. I told him our story. He was so impressed.

When they left, after being with us for ten days, I cried. I have never seen such a nicer group of men. Two weeks later we received a call from one of them. He was the man who made their arrangements for these trips. He said, "Carolyn, first of all thank you and Rob for a unforgettable trip. You treated us like family. Secondly, don't take any reservations for next summer. We came home and told our friends, families, and people we work with, and we have you booked for the summer." I was astonished, speechless, I heard, "Carolyn, are you there?" I was crying. He knew this and said, "I'll be talking to you soon."

Chapter 28

I never knew that the next time I talked to this man we would no longer own the Lodge. He came looking for us and was told we were working at the fish farm. The farm was across the water from the lodge. We looked across the water to Rendezvous Island. Our ten acres, our house, and our *Secluded Rendezvous Lodge*, no longer belonged to us. He drove up to the farm dock. He got out of his boat and we hugged, he and Rob shook hands. He looked at us and said, "I want to hear your story."

I began saying, "After I talked to you last summer we called Glen and Gary and told them your wonderful news about your filling us up with customers for the summer. Glen called us back in a couple of days. He said, 'You have to come in to the attorney's office to sign some papers. We need more money and have to take out a loan with the credit union.' Some of the people that helped build the new bath rooms, came to us and said they never got paid. When we told this to Glen he said he would take care of it. We thought Gary must have run out of money getting the "first class" look. This is what they called replacing everything. We agreed to take out a loan. We also thought Gary should not have to be paying for everything. I always thought Glen was using Gary. The day was set to see the lawyer, and we told Glen we were busy that day. He made the bookings, so he knew this. He told us the papers would be ready, just come in and sign. We drove the boat in. Glen met us in the Lobby. He said, 'It is all set, the loan has been approved, I just came from the credit union.' We three went in the office and sat at a round table. Glen took the paper signed it, and passed it to Gary. Gary took it and passed it to Rob, he passed it to the attorney and he gave it to me. Glen and Gary were talking loudly and laughing the whole time. I thought, they are sure nervous something is wrong! I looked at the paper. I asked him why it didn't say the name of the credit union, and why doesn't it say the amount and date, we are borrowing the money? The paper is blank. Gary yells, with a laugh, 'There she goes again, just sign it Carolyn.' I asked them if this was with Evergreen Credit. Glen said, 'yes.' The attorney said, 'I didn't fill it out because you were in a hurry to get back home.' I said, 'All you had to do was fill in the blanks.' Something is wrong I told myself, but it can't be wrong everyone signed it, and this is my attorney, too. He represents all four of us. I signed it." Why wasn't I listening? God or Jesus was telling me, no! I have never forgiven myself.

"The paper was for a Second Mortgage in someone's company number. I'm not allowed to say whom. It doesn't matter. God has taken care of them. The lodge had no customers; as you know, and we thank you, please thank them all for us. Because we could not come up with the money for the Second Mortgage, they took us to court. We talked to our attorney the night before the court. He said he would see us there. It was in another town. He did not come to court. The judge said, 'Your attorney didn't come. My hands are tied the Verdict goes to the Plaintiffs.' Rob and I looked at each other in disbelief. We could not believe this could happen this way. We were not allowed to say one word, it was done.

The judge gave us so many days to get out of the Lodge. Peter, who lived next door called us from Spain and said 'go stay in our house.' We did. We were so thankful to him. We were there a few months when a

Secluded Rendezvous

boat came up and a man standing in the boat called to us and said, 'come and work for me. My partner and I own a fish farm just across the water.' We said we would see him there tomorrow. We went over in the morning. His name was Jerry and he hired us. We started work right away. Rob would work with the fish and I was to cook and clean. We packed our things and left Rendezvous. And that is our story."

The man said, "I am so glad I found you and heard your story. I am so sorry and I speak for the other men too. You two worked so hard and took on such an unbelievable, beautiful task. God Bless, I hope we meet again."

Chapter 29

I loved my work. There were only five or six men in a shift to cook and clean for. They were all good to me. They spoiled me, in fact. We were there for one year. It was fun being with the fish. This was a very good farm, very clean, and no chemicals. Jerry loved his fish and made sure the men took good care of them, and they all did. One day, I was out with the men looking at the fish. Suddenly, my whole body was tired. I told Rob I was tired and was going to lay down upstairs. My whole left side and head felt strange and I said to myself, I am having a stroke. I picked up my flashlight, checked my eyes and body, and found my left arm was not moving properly. It felt numb and my fingers were bent like a claw. We flew out in the morning to the doctor. He told me I had a small stroke. We drove to Vancouver and my Cardiologist confirmed that I had a stroke. We went back to the farm. I felt fine, but my arm was still funny. One week later I was cleaning the seaweed out between two decks. It was raining and very wet. My left leg slipped straight back and my boot got caught on a shackle. My right leg slipped in between the two docks, up to my hip. I tried three times to pull myself up. Each time I let out a scream. Rob came and pulled me out. My knee was hurting. I went to stand on it and I fell. We went back to the hospital. I broke my tendon and tore three ligaments. I had to wear a cast until my surgery date, which was one year later. I took the cast off in three months. My leg was turning soft. I was fine, it just hurt a little, but I could walk. I never took a day off, but the partner told Jerry to let me go. Jerry said the partner was afraid we would sue him. I have never thought of suing anyone. It was not something I would do. I asked Robbie, "What we shall we do, where shall we go?" He said he was staying there. He had bought Randy's property next door to the Lodge. After crying all night, I woke up, and told myself, "You can and will do this—you have no other choice." I got up and packed and asked Rob to get me to Campbell River.

I left and lived two years more in Canada. I then decided to get a divorce. I needed to remain in Canada, the two years, in order to get my pension. Rob's cousin Elizabeth, asked me to live with her and Gordon. God bless, her. I stayed there a month, and then I moved to a small town called Tawissan. Robbie visited me often. I found a nice apartment and met two nice friends, Trinita and Diane. Diane and I went everywhere together. I was introduced to a couple, Bob and Lee. Bob was a doctor and Lee a nurse. They both were retired, and not too well and needed a gardener. I was hired. It was the perfect place for me to heal. These two people were lovely. They were so good to me. I could do anything I wanted in their Garden. Lee had worked in it for years, and it was beautifully done. I loved it. I also wrote a Cookbook, while I was there. Our good friends Tony and Mary helped me with this. I keep in touch with them, and Trinita, Diane, Bob, and Lee. They will remain in my heart always.

I moved back home to California, when my two years were up.

Robbie and I keep in touch, we are very good friends. I go to Canada to visit and Robbie and comes here from time to time. We never understood how two people could be married for years and never talk again. However, some people don't understand us either. We talk about maybe getting back together some

Secluded Rendezvous

year, but I can never leave my family again, and he has his family there. I suppose we will remain as we are. We are both happy where we live, and when we miss each other we make a trip to be together. He is building another house on the Island next door to the Lodge. He calls it *The Putter Without Pressure Project*. It's a place where the kids can all come and build a house if they chose. He is leaving it in care of his dear nephew, also named Robert Begg. His big dream is for it to be left as the Robert Begg Legacy. He even has a big building full of things for them to use in their places. We have been collecting items for a long time now for this. Robbie was living and working in Abbotsord for two years. He left his job and returned to the Island to stay for three months. He is building a house for the kids to have, when they come to build their own place. The house will have a foundation, floor, and roof completed, and a water tank with the system in this summer. Robbie and I talk at least five times a week. We tell each other how our day went. We also talk about us. He will be here the end of September. We are going to look for an old friend who has moved to Arizona. We are also driving to Mexico to visit some friends who invited us to stay for a couple of weeks. When we return, we will visit our families and friends here and in Sacramento. Robbie will be here three or four months. We will have lots of time together. He said, "Who knows, maybe I'll just stay with you." He didn't know, but I know—it's hard to live here in this different world after being on the island for twenty years.

    I spoke to Robbie one morning a week and a half ago. He was so excited in that he had just finished the last job he had planned on doing this summer. He said, "I only have one thing left to do, I am going to dive tomorrow and look for that anchor you and I built years ago and see if it's still there." I asked him not to dive alone. I reminded him that his equipment had not been checked. He said he had a friend there with him, who didn't dive, but this friend would have a rope tied to Robbie and the other end to himself, and the diving equipment was fine. I called Robbie the night of the dive several times and could not get an answer. I was so worried. He knew I was to call him at dinner.

I received a call, the next day, from young Rob who said, "Auntie Carolyn, I have some sad news, Uncle Robbie died yesterday while diving." His equipment failed him. The strong world that I had fit myself into now has some huge holes in it. I cried every night, all night long for two weeks. I took a look in the mirror and told myself, Robbie would not like me to do this. It has been two weeks now, and the tears are getting less and less. I know I will be fine again, because I know Robbie is with God. I know Jesus took him there. Robbie was a wonderful person who loved, and helped everyone. I shall miss my Buddy, my dear friend, and someone who I will hold within my Heart always. I feel very blessed for the memories and years, I have that we have shared.

I am now living in Mountain View, California. It is a lovely, friendly little town. I have a nice apartment. My son, Randall, his wife Linda, and my Grandson Josiah, live a few miles from me. They come over when I call, and are here for me when I need them. I am a Professional Photographer. I sold my Floral Designs for a few years. I paint, write, garden, and I have been taking lessons in Yoga for about two years. When I finish this book, I am going back to my Photography, and am extremely excited about this. For my birthday this year, Robbie bought me a digital Canon, a computer, and a printer. After being on the island all those years I have

Secluded Rendezvous

become a real loner, and it is okay. I do love people, but life is very different here. I feel well, happy, and blessed, and I know Robbie didn't leave me. He will be with me always. I have always tried not to think back on the things that were bad, that was yesterday. Each day is another day, and adventure for me. I am strong and I will do this…

Robbie

If not for you, I would not have written this book. I would not have known what it is like to live on an Island free and uncommitted. We planned our time day by day. When I was alone I chose where, when, and how I would begin my day. I was responsible for myself.

You and I lived our lives as such, for thirty-three years. My imagination carried me so far, I don't think I could have done this if not for you.

I know you are at peace now, with God, and for you I am happy. I am very strong and independent and thankful that I have learned how to be this, through the years.

I wake up each morning and say, "Thank you Lord for the wonderful sleep and am looking forward to a beautiful day." I am Blessed with my family and friends. What more could one wish for!

I carry you and our "Island Life" in my heart. Thank you my love, and dear friend.

Robbie using the Mobile Dimension Saw.

Logs to go under the lodge.

Carolyn Begg

Secluded Rendezvous Lodge.

Side view of the Lodge.

Secluded Rendezvous

Looking over docks onto Calm Channel.